Introduction to LANDSCAPE ARCHITECTURE

景观设计导论

伍晓雯 [美] 凯伦·尼尔森(Karen Nelson) —— 著

广西师范大学出版社
·桂林·

前言
景观设计教育探索

　　人们对人与环境以及优先事项的理解不断发展，景观设计学科也在随之转型。变化中的世界改变了人们对景观设计的需求和期望。三十年前的作品并不要求治愈地球，而如今获奖作品必须在生态上具有可持续性，才能被认为是美丽的。气候危机已经融入景观建筑领域，并促使人们通过参与景观设计来做出有意义且令人愉悦的事情。这本书展现了许多优秀的景观设计作品，每位读者都可以找到吸引他们的范例。

　　本书汇集了来自中国和美国的不同视角，鼓励初学者探索新的领域。作者们通过跨越大陆的共享历史的方式，描绘了对景观设计的整体感受，并导入了它在不同尺度上的设计发展路径。通过多种视角的探索，本书提出了建设"健康世界"的前进方向。本书的独特之处在于以案例研究突出技术革新和社会实践。从上海到巴黎，作者们将他们对城市的见解带入了不同日常生活和休闲活动场景中。

　　这本书包含对既定场地的分析性和客观性评估，同时融合了综合性和解释性的内容。这两种方式在设想当前和未来的景观设计时都发挥着重要作用。书中呈现了景观设计学科的不同特质。景观设计可能是充满想象的、理论性的和探索性的；也可能是受到实际情况和技术需求驱动的。书中自然地从在气候区中识别物种（植物和动物）过渡到感知空间体验的讨论。

　　这本书在许多方面帮助学生考虑未来实践的多种途径和模式。并提出了通过生态可持续的设计和互惠互利的规划来重新组织世界的方法。本书分享了设计师的见解，提出了景观设计专业学生可以思考的研究方向。作者们通过章节内容安排建构了一个潜在的学习视角，让学生了解农村和城市景观中需要考虑的元素。书中将城市作为重要的基础设施系统，帮助人们集聚，建立社交场所。同时，提出了水利基础设施和住宅花园对于城市的探讨价值。

这本书的核心理念是极具吸引力的——景观建筑正在变化，需要学生去理解并帮助塑造一个更具包容性的领域。不同章节作者的声音都增加了对景观设计的理解和研究角度。同时，每个章节又是独立的，探讨了不同条件或维度下的设计策略。所有章节共同形成了一个更大维度的视角：一个用于制作和阅读设计作品的视角。随着读者接触到更多的设计作品，他们返回每一章时，都会形成新的理解。

这本书提出了景观设计专业学生应当思考的问题，并展示了对这些问题的多种可能性的解答。作者分享了体验世界的愿景，以及在规划师、政策制定者、交通工程师和景观设计师定义的开放空间中游览和互动的方式。书中提出了城市如何自我更新以回应娱乐、避难、提供新鲜空气和降低温度的功能可能性。通过本书的学习，学生不仅会了解景观设计如何帮助居民社区更新发展，同时也能学会通过建议的数据库和搜索方式来进一步自我学习和研究。

本书鼓励学生和读者将景观设计视为日常生活的一部分，同时通过场地分析和项目设计来学习对地球生态的修复，以构建更具韧性的场所。

凯伦·尼尔森
美国波士顿建筑学院院长

Foreword

Exploration of Landscape Design Education

The discipline of landscape architecture transforms in parallel with evolving understandings of people, places, and priorities. The changing world alters the demands and expectations of landscape architecture. Works from thirty years ago were not expected to heal the planet. Recent award-winning work must be ecologically sound to be deemed beautiful. The climate crisis is now woven through the field of landscape architecture—and draws people to engage with the landscape to do something meaningful and gratifying. This book holds many outstanding works of landscape architecture. There is exemplary work that will engage each reader.

This book brings together perspectives from China and the United States to encourage a beginning reader to discover new territories. The authors give a sense of the shared history of landscape architecture across continents—and hint at new trajectories at many scales. Using several lenses to examine the field, this text suggests ways forward to make a healthier world. What makes this book distinctive is how it uses case studies to highlight technical practices as well as social ones. The authors bring their own insights to both daily and leisure behaviour from Shanghai to Paris.

This text incorporates both analytic and objective assessments of places as well as synthetic and interpretive ones. Both modes have a purpose in envisioning present and future landscapes. Some of these divergent qualities of the discipline of landscape architecture are manifested in this book. Landscape architecture may be imagined, theoretical, and speculative; landscape architecture may be governed by pragmatic and technical needs. The book seamlessly moves from identifying species (flora and fauna) in climate zones—to perceptual experiences.

In many ways, this book primes a student to consider the multiple paths and modes of practice that await them. The book also puts forward new ways to order the world to make it more resilient through ecologically sound and reciprocal design. This book shares insights and suggests questions that a student of landscape architecture can choose to cultivate. The book outlines a potential curriculum that identifies elements to be considered in rural and urban landscapes. The book holds up cities as the ultimate designed infrastructure to help people gather and create as part of a dynamic society. It identifies water infrastructure and residential gardens as works worthy of inquiry.

The fundamental premise of this work is compelling—that landscape architecture is changing—and needs students to appreciate and help forge a more inclusive field. Each chapter's individual author's voice adds to an understanding of and approach to landscape architecture. Each chapter isolates and explores a condition or dimension of the field. Together the chapters add up to something larger—a framework for making and reading work. Readers will return to each chapter with new understandings as they encounter more work.

This text identifies questions that a student of landscape architecture should ask and demonstrates divergent approaches to the possible questions. These authors share visions of experiencing the world and how to navigate and engage in open spaces defined by planners, policymakers, traffic engineers, and landscape architects. The text proposes possible responses to how cities can remake themselves to provide places of play, of refuge, of fresh air, and of cooler temperatures. The book shares modes in which landscape architects can help inhabitants thrive. It even helps students identify ways of researching similar projects by suggesting databases and search terms.

This book encourages students and readers to think about landscape architecture as part of daily life—and more significant aspirations to repair the Earth's ecosystems through careful analysis and design of projects that suit climate and vegetation to build in more resilient places.

Karen Nelson

Dean, School of Architecture at the Boston Architectural College

目 录
CONTENTS

CONTENTS

导言

Introduction

景观是各种自然过程的载体，这些过程支持生命的存在和延续，人类需求的满足是建立在健康的景观之上。

—— 美国景观设计师协会

The landscape is a vehicle for the various natural processes that support the existence and continuation of life, and the satisfaction of human needs is based on a healthy landscape.

——American Society of Landscape Architects

教材简介

　　《景观设计导论》是依托 2022 年上海市重点示范性课程《景观设计导论》而编写的环境设计专业基础课程教材。该课程是 2022 年上海市 50 门示范性全英语教学课程中唯一入选的设计类课程。课程以可持续景观为切入点进行专业教学，同时侧重设计实践在环境空间的运用。国内目前同类教材以理论分析偏多，且中英双语的设计基础教材建设较少，特别是在当代城市存量时代背景下，与城市更新、低碳社区改造、生态文明建设等国际设计议题结合的实践性内容欠缺。如何让学生在实践中探索景观设计的原理与可持续设计理念的形成的可能性，直观地感受景观设计实践中的设计方法、设计原理及设计材料，是建构设计思维的专业课程迫切需要解决的问题。新编双语实践型教材《景观设计导论》非常及时地为进入环境艺术基础课程的学生提供了更为广泛的思路与国际化设计创新的参考。

　　编写团队由理论与实践的多元专业人才组成，美国波士顿建筑学院师资团队也加入了编写及设计。本书采用中英双语的结构形式并配有大量的实践案例，目标在于让学生理解设计逻辑的形成，成为具有中西设计文化背景的实践型创新人才。

内容框架

　　本书分为五个部分。第一部分，即第一章，是景观设计的定义。第二部分为环境的认知，包括第二章和第三章。主要从自然环境和人工环境两个方面介绍环境的特征，以及环境组合的规律、准则和冲突问题。第三部分为尺度认知，包括第四章和第五章。这部分内容从人的尺度出发，阐述城市空间中对于不同尺度形成的空间感知，从而讨论尺度在场所人性化设计中的特征和设计原则。其中，第五章由美国波士顿建筑学院凯伦·尼尔森教授编写。第四部分为实践认知，包括第六章和第七章，主要介绍设计实践的调查方法、评价标准、规划准则的实际应用案例，展示设计研究的基本思路，开拓学生将基本理论转化为设计实践应用的方法。第五部分，即第八章，为责任和挑战，主要介绍当代环境面临的挑战和信息科技的创新，让学生理解其肩负的时代责任，具备应对新媒介时代的创新设计能力。本书以案例和设计主题带动理论知识的介绍，逐步引导学生在"做"的过程中形成设计研究思路和具体设计步骤。同时，为保证双语行文的流畅与清晰，本书未采用严格的逐字逐句中英对照，而力求学生在阅读不同语言版本时都能获得新的认知。依托本教材制作的"景观设计导论"线上课程已经在"学堂在线"平台开课，相关的课程课件、案例资料都可以在线上课堂查阅。

适用对象

如何让学生在有限的学期课时数内对景观设计专业有全面的认知，是编写本教材时重点考虑的问题。本书主要介绍景观设计与全球环境视角下的设计思考模式。通过多元的主题板块和设计训练，让学生将知识理论转化为对设计问题的思考，从而学会研究问题的本质，并理解设计背后的关系、逻辑与责任。通过章节的讨论训练，学生能够基本掌握景观设计的研究方法和思路，同时进入深层次的设计论证思维。本书也面向对景观设计感兴趣的非专业学生，特别是国际留学生，中英双语的形式让学生能快速检索相关的中外知识点。本书内容紧贴时代发展需求，强调科学和艺术共存的设计思维模式，为国际化市场需求培养卓越的设计创新型人才，开拓新的视角和课程设计方法。

Introduction

Background

Introduction to Landscape Architecture is a textbook for the undergraduate environmental design course based on the 2022 Shanghai Key Demonstration Course "Introduction to landscape architecture." The course is the only design course selected among the fifty courses of the 2022 Shanghai Demonstration in English, with the aim of creating an international course with Chinese characteristics. The course takes sustainable landscape as the starting point and focuses on the application of design practices in the environmental field. This is a bilingual and practical textbook that introduces the design principles of landscape design. Similar textbooks in China are mostly theoretical and there are almost no basic bilingual design textbooks. There is little practical content combining international design topics such as urban renewal, low-carbon community transformation and ecological civilization construction. In professional courses, it is important for students to explore the principles of landscape design and know how to develop sustainable design concepts, and for students to understand the feasibility analysis and exploration of design methods, principles, and materials. There is an urgent need to train design thinking through the practice of landscape design in the field. This book is timely, providing students entering undergraduate courses in environmental arts with a broad range of ideas and references for international design innovation.

The team involved in the preparation of the book is composed of professionals with diverse theoretical and practical knowledge, including the faculty from Boston Architectural College (BAC) in the United States. The bilingual structure and numerous practical examples are designed to enable students to understand the design logic and become practical innovators who understand the cultural background of Chinese and Western design.

Content Framework

The book is divided into five parts. The first part is the definition, written by Professor Maria Bellata of BAC, and includes chapter 1. The second part is environmental cognition, and comprises chapters 2 and 3. Here, the characteristics of the environment, both natural and artificial, as well as the laws, policies, and conflict issues of environmental combinations are mainly presented. The third part is the recognition of scale, covered in chapters 4 and 5. Starting from the human scale, we discuss the spatial perception created by different scales in urban space. Chapter 5 is written by Professor Karen Nelson of BAC, which will introduce the characteristics and design principles of scale for human-oriented design. The fourth part is practical perception and includes chapters 6 and 7, which mainly introduce the research methods, evaluation criteria, and practical use cases of design guidelines for design practice. The purpose is to understand the basic ideas of design research and to develop solutions to various problems that students face when implementing basic theories into design practice. The fifth part is responsibility and challenges. It focuses on the challenges of today's environment and innovations in information technology, so that students understand the responsibilities they must take on and develop the ability to respond to the new media era of innovative design. The introduction of case studies and design topics advance the introduction of theoretical knowledge and help students in the process of "making," to lead the design research ideas and specific designs, step by step. At the same time, to ensure the fluency and clarity of the bilingual text, this book does not adopt a strict word-for-word Chinese-English translation, aiming for students to gain new insights when reading different language versions.

Target Audience

How to provide students with a comprehensive knowledge of the landscape architecture profession in a limited number of semester hours is a central consideration of the author. This book focuses on landscape architecture and design thinking from a global environmental perspective. Through a variety of thematic panels and design training, students translate their knowledge and theories into thinking about design problems. In this way, they learn to examine the nature of the problem and understand the relationships, logic, and responsibility behind design. At the same time, this book is an introduction to design practice and is suitable as an introductory design teaching tool for environmental design, landscape architecture, urban planning, and architecture majors. Throughout the chapters, students are guided to construct design ideas and move into the depths of cognitive design reasoning, while beginning to master landscape design research methods and ideas. The book is also intended for non-students interested in landscape design. The bilingual format allows students to quickly access relevant Chinese and foreign knowledge and understand the process of putting design into practice through numerous case studies. The content of this book emphasizes the coexistence of scientific and design thinking, and opens up new perspectives and curriculum design methods for cultivating innovative design talents who can excel in the international market.

In summary, there is currently no bilingual textbook for introductory courses in landscape architecture in China. There are courses for higher levels in landscape design abroad, but less for introductory courses in landscape architecture. As demands for land are increasing, the development of courses on international topics such as urban renewal, low-carbon transformation of communities, and building an ecological civilization is weak. In response to the demand of the international market for innovative talent with an excellent education, bilingual design courses and teaching materials are needed to support talent development.

第一章

景观设计的定义及特点

Chapter 1

Landscape architecture and its characteristics

学习要点
Study Points

景观设计的定义和学科属性认知
Definition of landscape architecture and comprehension of disciplinary attributes

学习目的
Teaching Objectives

理解景观设计与自然因素、社会因素及经济因素之间的影响关系
To understand the relationship between landscape architecture and natural, social, and economic factors

景观设计专业的概念和定义，侧重于培养学生对景观设计的敏感性，并让学生作为环境的体验者进行自然环境与人工环境之间的对话。本章从景观设计的概念与范畴、学科属性、服务对象以及与全球可持续发展的关系进行介绍。学生通过理论知识和项目案例，理解影响景观设计的自然因素、社会因素及经济因素之间的关系。

This chapter aims to introduce the concept of "landscape architecture" and elucidate the essence of this profession. Simultaneously, it emphasizes the cultivation of students' sensitivity to the role of landscape architect, which will enable them to engage in a meaningful dialogue between natural and artificial environments, fostering a deeper environmental awareness. Within the conceptual framework of landscape architecture, various elements are introduced, and attributes are categorized into natural science, artistic, and service aspects. Through theoretical knowledge and project introductions, students can gain a comprehensive understanding of the relationship between natural, social, and economic factors that influence landscape architecture.

一、景观设计的概念与范畴

　　景观设计学的定义随着时代发展的需求在不断修正和拓展。国内学科体系中所称的"风景园林"专业在大部分情况下对应的是"landscape gardening"[1]，而本书范围内所对应的"景观设计"用"landscape architecture"[2]来表述更为贴切。那景观设计、风景园林以及环境艺术专业的区别究竟在哪里呢？景观设计学的定义是一个历史性概念，它会随着时代的发展和人们认识的提升而不断变化。[3][4]这也是为什么景观设计学的课程教材需要匹配时代更新的速度。自然环境的变化、城市危机，以及人们对环境品质需求的变化也在不断拓展景观设计的定义。对该定义的讨论及相关的学科知识扩展，对于学科的更新发展有着重要的推进作用。

　　从景观设计史的视角来看，景观设计的定义分为五个发展阶段。

　　第一，初起阶段。在有关英国绘画式园林风格[5]的书籍中，亨弗利·雷普敦和路登就使用过"景观设计"这一词组。但里面讨论的"景观"主要指的是"图像景观"，它只是对于风景的摄影捕捉，并不是真正意义上的景观设计师的工作。1858年，弗雷德里克·奥姆斯特德[6]和卡尔弗·沃克斯[7]在美国纽约中央公园设计竞赛中第一次以景观设计师的身份胜出。当时的参赛者以建筑师为主，以景观作为主要考虑对象进行城市空间设计，呼应了当时城市发展的变化。在城市发展的进程中，我们同样需要关注自然环境。1863年，纽约中央公园委员会第一次正式使用"景观设计"这一术语作为一种新的职业。

1.1 The concepts and categories of landscape architecture

The definition of landscape architecture is continuously evolving and expanding to meet the changing needs of our times. While the term "landscape gardening"[1] typically pertains to planting on a specific scale, "landscape architecture"[2] serves as the more appropriate focus of this book. What sets landscape architecture apart from environmental art and landscape gardening? Landscape architecture is a concept rooted in history, undergoing continuous evolution alongside changing needs and ideas.[3][4] As such, course materials concerning landscape architecture also evolve. Transformations in the natural environment, urban crisis, and growing demands for environmental quality contribute to the ongoing expansion of the definition of landscape architecture. Delving into this definition and expanding related subject knowledge stands as a vital contribution to the rejuvenation of this field.

Looking back at the history of landscape architecture, its definition can be divided into five distinctive development stages. The earliest phase can be attributed to figures such as Humphrey Repton (1752–1818) and JC Louden (1783–1843), who were referred to in books about English painting styles as "landscape architecture." [5] However, the term "scenery" in this context primarily refers to "picturesque scenery" and represents a photographic capture of the landscape rather than the work of a genuine landscape architect. We can trace the beginning of the first phase to 1858 when Frederick Law Olmsted (1822–1903) [6] and Calvert Vaux (1824–1895) [7] won the competition for landscape architects for the design of Central Park in New York. For contestants, who were primarily architects at that time, giving paramount consideration to the landscape in urban space design also aligned with the shifts occurring in urban development during that era. In the course of urban development, greater emphasis on the natural environment became imperative. It was in this way, too, that the term was formally recognized as a new profession in 1863 by the New York Central Park Commission.

During the second stage, design language took on a more scientific character, heavily influenced by the industrial style of the Bauhaus School in the nineteenth century.

第二阶段，在19世纪包豪斯工业风格的影响下，设计语言变得更讲究科学原则。从建筑设计到景观设计，无论是平面造型，还是人体工学的尺度，都更趋向于极简风格。在建筑及人工环境的极简趋势下，景观元素以保持原有自然状态为主，多与建筑语言进行对比。在路德维希·密斯·凡德罗[8]的"玻璃盒子"中，建筑幕墙以全玻璃的造型出现，除了呼应工业时代的材料和极简主义风格外，还传达了一个重要的理念：建筑以一种隐形的媒介出现，而人的体验与周围的景观环境才是真正的中心。

第三阶段强调协调。克里斯托弗·唐纳德[9]在《现代景观中的花园》[10]一书中提出，景观设计就是自然的协调者。它在城市空间设计中具有重要作用，协调着自然环境和城市发展需求、政策秩序、居民体验、建筑设计、交通管理等各种城市问题的关系。从这一阶段起，景观设计脱离"花园"的功能类型，涉及内容极其广泛，与城市生活的各个方面形成密不可分的关系。

第四，高速发展阶段。当城市化范围不断扩大，景观设计对于城市整体的规划设计以及未来的体验发展起着重要的定调作用。20世纪，巴黎在"二战"后进行的城市规划更新，就以"花园城市"[11]作为总体的规划设计基调，以城市中央的广场或者标志性建筑为中心，进行林荫道的布局（图1-01）。街道的尺度以人居尺度为主，注重人车分流以及城市总体的天际线控制。以景观设计为核心理念的城市规划打破了人们"景观设计只是绿化设计"的认知边界。

第五，生态应用思考阶段。随着存量时代

A noticeable inclination towards minimalism emerged, evident in both the two-dimensional form and the ergonomic scale. Within the backdrop of this minimalist architecture trend, the landscape elements predominantly retained their original natural state, serving as a point of comparison to architectural language. For instance, the all-glass curtain wall created by Ludwig Mies Van der Rohe[8] mirrored the materials and minimalist aesthetics of the Industrial Age, while also conveying a profound concept. Architecture, as an intangible medium, emphasized the pivotal role of human experience and the surrounding landscape environment as the true focal points.

The third stage emphasizes coordination. Christopher Tunnard[9], in his book *Gardens* [10], proposes that landscape architecture serves as the coordinator of nature (to plant is but a part of landscape composition; to co-ordinate is all). In the design of urban spaces, landscape architecture assumes a crucial role in harmonizing the natural environment, the demands of urban development, policy enforcement, orderliness, residents' experience, architectural elements, traffic management, and other urban challenges. When landscape architecture transcends its functional association with "garden," it assumes a more expansive role, intimately interconnected with all facets of urban life.

The fourth stage represents a period of rapid development. As urbanization continues to advance, landscape architecture assumes a pivotal role in setting the tone for overall urban planning and design, thereby shaping the future of the urban experience. In the twentieth century, following World War II, Paris underwent a rapid transformation with the implementation of a master plan for a "garden" city[11] (Fig 1-01). This plan featured a layout centered around boulevards radiating from the central square or a prominent city landmark. Street design primarily adopted an artificial scale, prioritizing the separation of pedestrians and vehicles, and the regulation of the city's overall skyline. Urban planning rooted in this concept of landscape architecture transcended the conventional notion that it solely entailed green design, expanding its scope and influence.

的发展，城市空间更注重质量和未来城市的可持续应用。景观设计从大型城市开发建设转向城市绿地的优化更新。在城市更新中景观设计更倾向于"微"建造，"深"影响。[12] "微"建造，代表景观设计作为软连接，在公共绿地中对于城市生活活动的事件营造比功能建造更为重要。"深"影响，代表在城市绿地建设的过程中，更为注重自然环境、人工环境、人文环境与城市资源的多元连接，而非单一化绿地开发。景观设计的概念与定义在时代的发展中不断被拓宽，包含了自然和人工环境共同生存的无限可能。

二、 自然科学与艺术属性

景观设计的自然科学属性，可归纳为大自然中有机或者无机的事物和现象，包括天文、生物、地理等因素。在景观设计的基本构成中，可分为软质元素和硬质元素。这两类元素都来自自然科学。其中，软质元素可归纳为树木、水体、风向、雨水、阳光、天空等[13]；硬质元素可归纳为铺装、墙体、栏杆、座椅、凉亭等构造物。

有同学可能会问，景观设计中的自然科学属性分类与环境科学的学科分类非常相似，都是和自然环境打交道的学科，那么两者的区别在哪里？这里便需要提及景观设计的艺术属性和社会服务属性。涉及社会服务属性的内容较多，我们在下一小节单独介绍。从艺术属性上分类，景观设计涉及的自然科学元素并不完全是它自然的状态，设计的需求会改变它的自然形态。例如，在中国的古典园林中，移植进庭院的花草植物和石头通过艺术化的组合再现自

The fifth stage signifies the era of applied ecological thinking. In the era of stock competition, marked by intense competition for resources, the development of urban spaces places greater emphasis on quality and sustainable urban applications. Landscape architecture has transitioned from predominantly large-scale urban development and construction projects to the optimization of urban green spaces, fostering the rejuvenation of urban public spaces. The optimization of landscape architecture in the context of urban renewal often involves a more "microscopic" approach, but its impact runs deep.[12] This "micro" approach embodies a delicate connection between landscape architecture and urban green spaces. It prioritizes the creation of events for urban life activities in public green spaces and urban public areas over physical constructions. The profound influence of this approach lies in its design of green spaces and urban development, emphasizing the intricate dynamic among natural and artificial environments, human habitats, and urban resources rather than focusing solely on the development of individual green spaces. Over time, the concept and definition of landscape architecture have expanded to include the coexistence of natural and artificial environments.

1.2 Natural science and artistic attributes

The natural science attributes of landscape architecture encompass the organic and inorganic elements and phenomena found in nature, including astronomical, biological, and geographical factors. The fundamental components of landscape architecture can be categorized as either soft elements or hard elements, both of which have their roots in the natural sciences. Soft elements comprise entities such as trees, water, wind, rain, sunshine, and the sky, among other things.[13] In contrast, hard elements consist of features like pavements, walls, railings, seating arrangements, pavilions, and various other structures.

Some students might argue that the classification of natural attributes in landscape architecture closely resembles that used in environmental science. Both fields are linked to the natural environment, so what sets them apart? In addition, it's essential to acknowledge the artistic attributes and

然的浓缩之美。植物和石头的自然属性没有变化，但是重组后形成的新的空间形式，以及它带来的艺术价值和社会价值才是景观设计中重要的部分。

在西方园林史上，对自然元素进行艺术化处理的例子更为鲜明，从文艺复兴的朗特庄园到近现代公园的代表纽约中央公园，都体现着自然植物和自然形态在景观设计范畴下的艺术转化，成为同时期园林设计的典范以及各自地域的代表性景观。[14] 此外，除了光照、水体、植物等自然属性的影响，人造环境强烈的艺术属性也影响着景观设计。以安迪·高兹沃斯[15]为代表，通过人工模仿自然，形成独特的景观场所，让平凡的日常场景具有特殊的艺术效果（图1-02）。美籍华裔建筑设计师林璎[16]，通过人工建造的手法模仿自然地形而形成建筑和景观设计（图1-03），不仅在手法上体现出对于自然科学属性的模仿，在艺术表达上更是把自然科学属性中的美放大成为我们可以体验的场所。她设计的越战纪念碑中的下凹地形，通过场地的高低变化形成特殊的瞻仰怀念路径，并在四季的自然环境中展现出对于战争的无声抗议——一种无法磨灭的伤痕一直留在人们的心中。印度裔英国雕塑家安尼施·卡普尔[17]所创作的雕塑《云门》如同建筑的体量，通过镜面反射周边的城市景观，观察者通过雕塑的反射来体验都市景观和文化气息（图1-04）。景观设计的元素具有强烈的自然属性，进行景观设计时，区别于其他艺术创作，一方面我们需要尊重其自然属性；另一方面，景观设计思考的过程中，我们也需要具备艺术属性的想象力，在自然和人工技术的限制下，对城

societal contributions of landscape architecture. Given the significant involvement of social attributes, we will address them in the subsequent section. Regarding the classification of artistic attributes, it's worth noting that the natural science elements mentioned in landscape architecture are not completely left in their natural state but undergo transformation based on design requirements. For example, in classical Chinese gardens, flowers, plants, and stones transplanted into the courtyard maintain their natural essence while expressing their beauty in concentrated and artistic combinations. The natural attributes of these plants and stones remain unchanged, but the new spatial forms crafted through rearrangement, along with their artistic and social significance, constitute an integral aspect of landscape architecture.

For artistic purposes, the extensive utilization of natural elements is even more evident in the history of Western gardens. From Lange Manor during the Renaissance to contemporary examples like New York Central Park, which represents modern parks, they all showcase the artistic transformation of natural conditions through the strategic placement of plants and the construction of natural forms. These endeavors establish the artistic paradigm of their respective era and evolve into distinctive landscapes.[14] On the other hand, in addition to the impact of natural attributes such as light, water, and plants, artificial environments also possess strong artistic characteristics that exert influence on landscape architecture. Under the leadership of public artist Andy Goldsworthy[15], unique landscape spaces have been fashioned by artificially emulating nature, thereby imparting special artistic effects to ordinary daily scenes. (Fig 1-02) American Chinese architect Maya Lin[16] employs artificial building techniques to mimic natural terrain in her architecture and landscape projects. (Fig 1-03) Her designs not only articulate scientific attributes but also magnify the inherent beauty of natural attributes of places, allowing for artistic expression and creation to flourish. For instance, the sunken topography of the Vietnam War Memorial in Washington carves out a special path for reflection and remembrance, its shifting elevations within the site serving as a silent protest against the war within the natural environment—a lasting emotional imprint etched in people's hearts. Furthermore,

市的未来空间进行富有想象力的创作。

三、社会服务属性

景观设计面对的设计对象范围广泛并且影响深远。地理学中的"景观"作为科学研究的对象，指的是"反映统一的自然空间、社会经济空间组成要素总体特征的集合体和空间体系，包括自然景观、经济景观、文化景观"。[18] 当我们思考如何进行景观设计的时候，首先要关注的是为谁服务，以及设计将会带来怎样的影响。

景观设计对于自然保护、经济促进以及文化教育都具有重要的服务性影响。[19] 从宏观的角度出发，它涉及自然环境中的绿地、林地、湿地、水体和山地，同时也包括人类环境与动物环境的栖息关系。面对自然环境的改造设计，需要注意人居环境对动植物生态环境的影响。例如，城市扩张中，公路基建会加剧自然林地破坏以及周边污染，因此设计师在进行整体环境规划的时候不仅需要考虑设计对象，还需要思考其设计活动对未来自然资源和城市发展的影响。此外，景观设计主要服务对象涉及工程的落地。在设计创意转化为真实项目的过程中，施工技术、材料选择以及建造成本预算都是设计师需要考虑的问题。这也同时赋予景观设计更大的社会责任，应考虑到设计过程中对自然和社会环境的影响。例如，美国纽约高线公园[20]的选址本是一段即将被拆除的破旧铁路，但是设计师通过对铁轨进行工业化设计语言的转换，使高线公园成为纽约最有活力的公共空间之一。[21] 并且因为高线公园改建的成功，周边原本荒废的厂房也重组业态，形成集创意

British-Indian sculptor Anish Kapoor's *Cloud Gate*[17], also known as "the Bean" in Chicago, establishes a spatial threshold within a new park. His bean sculptures, akin to architectural designs, reflect the surrounding urban landscape through mirrored surfaces. Observers can experience the city's landscape and culture through the sculpture's reflective prism. (Fig 1-04) The elements constituting landscape architecture inherently possess pronounced natural attributes. Throughout the creation of landscape architecture, unlike various other forms of artistic creation, it is imperative to begin by respecting the inherent properties of the natural environment. Simultaneously, during the contemplation of landscape architecture, an imaginative approach is essential, allowing for the crafting of visionary spaces for the future of the city within the constraints of both natural and artificial technologies.

1.3 Social service attributes

Landscape architecture involves a diverse array of design objects, yielding far-reaching influences. In the realm of scientific research, the landscape, as defined in geography, represents "a collection and spatial system reflecting the overall characteristics of the components of a unified natural space and socio-economic space. It includes natural scenery, economic scenery, and cultural scenery."[18] When contemplating the design process, our foremost consideration should be identifying the intended audience and assessing the impacts of our design decisions.

Landscape architecture plays a crucial role in preserving nature and natural systems. It also contributes to economic development, culture, and education.[19] From a macro perspective, landscape architecture intentionally creates the natural environment, encompassing aspects such as forests, green spaces, woodlands, wetlands, water bodies, and mountains. It also designs the intricate relationship between human habitats and the habitats of other creatures within this environment. When engaging in the transformation and design of the natural environment, it is imperative to consider the influence of human settlements on animal and plant ecologies. For instance, the expansion of cities and the construction of road infrastructures can exacerbate the destruction of natural forest lands and escalate the pollution in their vicinity.

园区、商业零售和艺术展览等多元活动为一体的社区中心。高线公园的改造体现了景观设计对于公共绿地的设计营造方式和对城市片区整体经济活力的影响。

总体说来，景观设计是展现区域人文特征的重要手段。景观设计包括了城市公共空间中的各种休闲活动场地，如公园绿地、街道景观、滨水景观、屋顶花园以及儿童游乐场地等。每个设计类型都是不同的社交、文化和艺术形态的传播平台，同时也体现了城市特别的精神文化元素。中国古典园林中对于"景观场所"的定义就有"微缩的宇宙"的含义。它通过"通""流""隔""抑""曲""烟"等造园手法诠释物质世界与精神世界的微缩统一[22]（图1-05）。在古人的造园理念下，自然景观和人工景观都被看作生命体，不仅是物质的

表现，更是思想灵魂的代表。景观设计在设计对象和社会服务方面任重而道远，这要求设计师在思考设计方案时需要有更广阔、更全面的考虑。

四、人居环境与全球可持续性发展

"可持续性"这个词在定义上，为"保持，支持或坚持"。从1987年3月开始，联合国布伦特兰委员会提出："可持续发展就是既满足当代人的需求，又不损害后代人满足其需求的能力的发展。"随着环境资源的消耗、人口数量的增长以及污染程度的加剧，对以可持续发展为目标，以保护地球的人居和自然环境为目标的设计的需求不断增加。景观设计是人居环境和自然环境协调发展的重要媒介，这便要求景观设计在构思过程和实践探索中始终贯穿

When planning for the overall environment, designers should take into account not only human activities but also the long-term influences of their design on natural resources and urban development. Furthermore, landscape architecture involves the practical implementation of projects, necessitating the translation of design concepts into tangible outcomes. This process includes decisions regarding construction techniques, material selection, and estimation of construction cost. This also brings greater social responsibility to landscape architecture, as they must consider the influence of the design process on both the natural and social environment. Take, for example, the transformation of the High Line[20], an elevated linear park in New York City. Through the repurposing of a disused railway track, it has evolved into one of the city's most vibrant public spaces, incorporating industrial design elements.[21] This transformation has not only revitalized the neighboring abandoned factory buildings but also reimagined them as a diverse community center, featuring design and creative parks, retail spaces, and art exhibitions. The metamorphosis of the High Line not only reflects the spatial design of a public green space area but also demonstrates

its far-reaching impact on the overall economic vitality of the urban area.

In conclusion, landscape architecture is an important means of expressing the unique human character of a region. It includes a multitude of leisure activities within urban public spaces, such as parks, street scenes, waterfront areas, roof gardens, and children's playgrounds. Each variant of landscape architecture serves as a platform for the communication of diverse social, cultural, and artistic forms, collectively contributing to the fabric of a city that reflects specific cultural and spiritual elements. The classic definition of a "landscape place" in traditional Chinese gardens carries the profound meaning of a "microcosm of the universe." Through gardening techniques such as "passing," "flowing," "separating," "suppressing," and "bending" (Fig 1-05) [22], It unites the material and spiritual worlds within a microcosm. In the ancient gardening philosophies, both natural and artificial landscapes were perceived as living entities, not merely as objects of expression but also as embodiments of the soul and the mind. Landscape architecture assumes a pivotal

图 1-01　**Paris urban planning**
　　　　巴黎城市规划

图 1-02　*Storm King Wall* **by Andy Goldsworthy**
　　　　《风暴王墙》，安迪 · 高兹沃斯

图 1-03　*The Takeaway* **by Maya Lin**
　　　　《2×4 风景》，林璎

图 1-04
Cloud Gate sculpture in Chicago, USA
美国芝加哥《云门》雕塑

图 1-05
Lion grove garden in Suzhou, China
狮子林，中国苏州

可持续发展的理念，为未来健康的环境资源提供更有效的设计思路。这些举措应与消除贫穷的一系列战略合作并进，包括促进经济增长，解决教育、卫生、社会保护和就业机会的社会需求，遏制气候变化和保护环境等。

作为可持续发展的组成部分，景观设计的方法论需要关注四个方面的变化，包括时间维度、人口模型、自然资源和社会需求。[23] 从宏观层面思考，人类文明的进化与地球地质史有显著关系。从地质学中最开始的地球的形成、生命的起源、前人类时代、直立人时代、现代人、文明的起源到当代的城市环境，环境景观始终伴随着地球文明的发展而改变。地球地质景观的发展与人类活动和设计建造相互影响。时间维度让我们思考设计应顺应自然规律，人类和地球是繁荣共同体，二者密不可分。地球

上人类的总人口一直占据生物层面的顶层，同时医疗技术的发展也带来寿命的延长。自从智人诞生后，人类不断地繁衍后代，截至21世纪20年代，总人口已有80多亿。人口数量增长导致自然资源过度使用，也导致对其他物种生存环境的破坏（图1-06、1-07）。人类对可再生资源的消耗包括过度捕捉鱼类、树木砍伐、废气排放和土地开发等，这直接导致空气、土地和水资源的污染，对大气层的破坏继而引发全球变暖和其他自然灾害。为了实现人居环境可持续发展，人类需要控制过度的能源消耗和污染的蔓延，调整城市发展的方式和方法。营造既能满足人类后续发展的需求，又能平衡自然生态发展的空间，是当代设计师必须承担的责任。[24]

and far-reaching role within the realm of design and social service responsibility. This demands that designers adopt a broader and more comprehensive perspective when formulating design schemes.

1.4 Human habitat environment and global sustainable development

According to its definition, "sustainable development" entails maintaining, supporting, or persisting. In March 1987, the United Nations Brundtland Commission proposed: "Sustainable development is about meeting the needs of the present without compromising the ability of future generations to meet their needs." As environmental resources deplete, populations expand, and pollution spreads, the demand for sustainable design to safeguard the Earth's natural environment grows. Landscape architecture served as an important medium conduit for fostering harmonious development between humanity and the natural world. This requires the integration of sustainable design considerations into every stage of the conceptualization process and practical exploration within landscape architecture, offering more

effective design methodologies that can contribute to a healthier environment for future generations. These initiatives work hand-in-hand with broader strategies for poverty alleviation, including promoting economic growth, addressing social needs such as education, healthcare, social welfare, and employment opportunities, mitigating climate change, and protecting the environment.

As an integral component of sustainable development, it is crucial to consider four dimensions of change when thinking about landscape architecture methodologies: the time dimension, population modeling, natural resources, and social needs.[23] Within the time dimension, when examined from a macro perspective, a profound connection exists between the Earth's geological history and the evolution of human civilization. Geologically, the Earth has witnessed transformations from its inception to the emergence of modern humans, culminating in the evolution of contemporary urban environments. Throughout this temporal journey, the environmental landscape has evolved in tandem with the progression of civilizations. From this perspective of time, we should

图 1-06 **Desertification of land**
土地荒漠化

图 1-07 **Air pollution**
大气污染

ensure that every design adheres to the principles of natural laws, acknowledging that humans and the Earth form a prosperous and interdependent community. The development of Earth's geological landscape deeply influences the planning and design of human activities. Human beings have perpetually occupied the apex of the biological hierarchy. At the same time, advances in technology and medicine have enabled population growth and increased longevity. Over the past 200,000 years, the human population has expanded, currently exceeding 8 billion individuals. However, such growth of the population has led to the overexploitation of natural resources, resulting in habitat destruction for other species (Fig 1-06, 1-07). Human activities, including overfishing, deforestation, fossil fuel consumption, and land development have led to the depletion of renewable resources and contributed directly to air, land, and water pollution, further exacerbating issues such as climate change and other natural disasters. To achieve sustainable habitat development, human societies must curtail excessive energy consumption and the diffusion of pollutants. Consequently, it becomes imperative to adjust the methods

and approaches to urban development. Contemporary designers bear the responsibility of shaping spaces that cater to the long-term needs of future generations while also striking a balance between the preservation and development of natural ecosystems.[24]

1.5 Explore landscape architecture under the sustainability concept

This is indeed a global issue of the times, presenting a significant challenge that contemporary landscape architects must confront. Within this context, there are four development principles that landscape architecture should follow.

1.5.1 Balance ecology, economy, and aesthetics

The concept of sustainable development emphasizes the need to consider recycling and regeneration in the development of resources, materials, and economies. Landscape architecture is a design discipline that integrates ecology, economy, and aesthetics, and at the same time, it requires a design strategy for resource development and recycling throughout the design process. Key design

五、如何在可持续理念下探索景观设计

这是当代景观设计师需要面对的全球时代议题和挑战。探索可持续理念下的景观设计需要遵从四个发展规则。

1. 生态、经济与美学的平衡

可持续概念意味着在资源、材料和经济方面都要关注循环再生长过程。而景观设计是集生态、经济与美学为一体的设计学科，这要求我们在探索实践时，必须具备资源开发和循环利用的设计策略。自然环境中的水资源、硬质铺装、建筑构架、植物营造等都是景观设计重要的对象，对其进行适应区域开发，对保护原有自然生态系统有着重要意义。铺设草坪是景观设计中常用的种植手法，但如果在干旱地区采用大面积草坪种植，就需要消耗大量水资源进行灌溉，那么这是一个好的设计吗？因此，设计前期需要进行生态、经济、美学三方面的平衡思考，通过创新应用，寻找平衡这三个方面的手法。例如，绿植和草坪是景观设计中常用的造景元素，同时需要大量灌溉，因而，通过给绿植和草坪增加雨水的渗透量并减少地表面径流，是较为常见的可持续设计方法。另外，也可以通过在大型公共空间场所设立雨水收集系统，铺设透水铺装和再循环利用灰水，对场地的绿植和草坪进行灌溉。景观设计通过城市空间的雨水系统规划、绿植设计和铺装设计，可有效节约水资源，实现可持续性。类似的设计方法不断被应用在不同的国家和项目中，随着材料技术的发展，新的方法也在不断产生。我们希望学生在学习景观设计基础的时候，就能以环境共同体的视角，从生态、经济和

elements in landscape architecture encompass water resources in the natural environment, hard pavement, building structures, and plantings. It is essential to adapt these elements to regional development while protecting the primitive natural ecosystem. Laying lawns is a common practice in landscape architecture. However, when lawns are planted in extensive arid areas, irrigation consumes excessive water. Therefore, it raises the question of whether this is a prudent design choice. Prior to embarking on design, careful consideration of the balance between ecology, economy, and aesthetics is necessary. Through the application of innovative design methods, we endeavor to balance these three aspects of design expertise. For example, greenery and lawns are common landscape elements that demand substantial plant irrigation. An established sustainable design practice in landscape architecture involves enhancing rainwater infiltration and mitigating surface runoff through the incorporation of greenery and lawns. Alternatively, rainwater can be collected, facilitating permeable paving and greywater recycling in large public places through rainwater collection systems. These collected resources can then be employed to irrigate green spaces and lawns. Through meticulous planning, greenery integration, and pavement design of rainwater systems in urban areas, landscape architecture can effectively conserve water resources, achieving a more sustainable landscape. Such sustainable landscape architecture methods are consistently employed across diverse countries and projects. This approach to sustainable landscape architecture evolves alongside advancements in technology and material sciences, resulting in a plethora of innovative design techniques. We aspire for students to comprehend the fundamental principles of landscape architecture while contemplating the future practice of sustainable landscape architecture through the lens of environmental stewardship, economic prudence, and aesthetic perspectives.[25]

1.5.2 Maximizing resource recovery and use utilization
Landscape architecture involves the utilization of diverse materials, requiring substantial manpower and financial resources throughout the phases of development, construction, planting, and maintenance. Therefore, to

美学的角度共同思考未来可持续景观设计的实践方法。[25]

2. 资源回收和利用最大化

景观设计涉及较多材料的运用。从场地开发、工程建造、植物种植到后期维护都需要消耗大量人力和财力，所以设计时需要考虑每个阶段的材料回收与利用情况，从而形成最大化的可持续利用。

在场地开发过程中，对已有场地进行改造比兴建全新的场地更能提升材料回收率。已有场地的改造，可以对场地原有的结构和景观资源进行再匹配，从而减少树木砍伐和运输的成本。在工程建造中，如果使用当地的材料施工，同样可以减少材料的运输成本，并且建造过程中的沙石、种植土和木料都可以通过地形的挖空和堆叠形成自然填补，最大化地利用场地的资源条件。植物种植方面，选择本土植物比移植植物更容易降低维护成本。本土植物对当地的气候条件具有较强的适应性，不需要过多的养护，但是如果移植跨地区的植物，可能会产生更多的光照培育或者灌溉需求。对于大面积的绿地种植而言，这不符合可持续景观设计的理念。

3. 尊重自然生态格局与生态功能，减少对自然环境的干预

学生在学习景观设计基础时，需要了解环境共同体的概念，目的是学习尊重。尊重自然环境，尊重人居需求，尊重发展与平衡。景观设计在场地开发准备中，需要了解当地的自然生态环境，包括自然植物、水体资源、土壤的酸性、动物的栖息偏好等，并通过调查形成较为完整的场地认知。景观设计须基于这个认知

achieve sustainability, it is important to prioritize material recycling and utilization at each stage of the design process.

Opting for site transformation rather than starting with a fully developed site during the site development process can significantly improve material recovery rates. The redevelopment of existing sites allows for the harmonization of the original structure and landscape resources of the site, thereby reducing the need for tree felling and transportation, consequently curbing costs. Moreover, employing local materials in project construction will further reduce material transportation expenses. Techniques such as hollowing out and stacking can be used to maximize the use of the site resources, including sand, gravel, planting soil, and wood, to contour the terrain effectively. In terms of planting, choosing indigenous plants for design offers a pragmatic approach to reducing maintenance costs compared to transplanting plants. Local plants are highly adaptable and can thrive within the local climate and water conditions, requiring minimal maintenance. However, introducing non-native, trans-regional plants could lead to increased artificial light cultivation and water irrigation. Such practices are at odds with the principles of sustainable landscape architecture that promote extensive green planting practices.

1.5.3 Respecting natural ecological patterns and functions and minimizing interference with the natural environment

As the cornerstone of landscape architecture, it is essential for students to grasp the concept of environmental community. This entails learning to respect the natural environment, comprehend the requirements of human habitation, and development, and strive for balance. When preparing for site development, landscape architects must delve into the intricacies of the local natural ecology. This encompasses an understanding of native plants, water resources, soil acidity, animal habitats, etc. By conducting a thorough investigation of the natural environment, a comprehensive understanding of the site can be attained. Only through a wealth of knowledge and sincere respect for these natural elements can we effectively orchestrate the development and balance between human activities and the environment.

来协调自然环境与人居环境，只有在具备足够的认知和尊重的前提下，才能兼顾两者的发展与平衡。

4.通过科技与设计的跨专业合作进行城市的可持续设计

当代景观设计师的责任已经不仅仅是服务于植物园艺。数字技术与国土空间规划的发展，也为景观设计提供了可持续设计的新手段。秉承可持续设计的理念，绿色建筑以及绿色规划标准已经在全球设计背景下不断完善。其中以美国LEED[26]的绿色建筑标准规范推行较多，英国和中国也展开了绿色建筑技术研发。景观设计与规划的可持续设计方法，具体体现在节能减耗、提升空间环境品质、提高人体舒适性、保护生态环境等方面。在节能减耗的过程中，注重使用可循环使用或者可再生材料。例如，

可关注具有FSC[27]认可的木材机构，在木材的砍伐、加工、再种植的过程中都能达到零碳循环的工厂。此外，减少水、电、光等能源的运用，如建立环境光感应数据调节照明能耗，利用水冷热泵以及地暖泵双重调节空调能耗，都是减少日常建筑空间能源消耗，同时最大化提升人体舒适感的措施。景观设计的能源节约还体现在雨水的循环和多向利用上，如雨水回用。滴灌系统能更好地保持土壤中水、空气、热度等的良好状态，同时具有节水、省工、节能、节肥的优点，比传统的灌溉方式节约60%的电力，水利用率可达97%。[28]

1.5.4 Leveraging technology and design for sustainable urban design

The role of the contemporary landscape architect extends beyond conventional garden design. The emergence of digital technology and spatial land planning has ushered in a new era for sustainable landscape architecture. Aligned with the principles of sustainable design, global standards and specifications for green buildings and planning standards have undergone refinement. In the United States, LEED[26] (Leadership in Energy and Environmental Design) green building standards and specifications have seen increased adoption, while the United Kingdom has also started the research and development of green technologies within China. The sustainable design methods inherent in landscape architecture and planning revolve around reducing energy consumption, enhancing the quality of the landscape environment, improving human comfort, and preserving the ecological environment. In the pursuit of saving energy and reducing consumption, considerable emphasis is placed on using building materials that are either recycled or reclaimed. For instance, a timber building certified by the FSC[27] (Forest Stewardship

Council) can establish a carbon-neutral cycle spanning the entire wood lifecycle, from harvesting and processing to reforestation. Simultaneously, it reduces the use of water, electricity, and lighting in manufacturing processes. Energy consumption for lighting is managed through ambient light-sensing data, and the energy consumption of air-conditioning is double-regulated via water-cooled heat pumps and underfloor heating pumps. These strategies effectively reduce energy consumption within building spaces while optimizing human comfort.

Energy saving in landscape architecture predominantly manifests through the recovery and utilization of rainwater. The specific process is as follows: roof and ground rainwater is filtered and disinfected before feeding into rainwater purification tanks. The deployment of drip irrigation systems facilitates the maintenance of soil quality, offering advantages such as water and labor saving, as well as reduced energy and fertilizer usage. Compared to traditional irrigation methods, drip irrigation can achieve significant savings, conserving up to 60 percent of electricity and a remarkable 97 percent of water resources.[28]

小结

　　在景观设计实践过程中，应将可持续设计的理念融入从前期设计到施工维护的全过程，从而让设计项目达到真正的节能低耗，让人居环境和自然环境通过景观设计协调共存发展。

Summary

Modern energy-efficient technologies and the use of low-carbon materials are instrumental in implementing a variety of sustainable design methods in landscape architecture. In practice, it is essential to incorporate the principles of sustainable design across the entire lifecycle of landscape architecture projects, spanning pre-design, design discussions, construction, and maintenance. This holistic approach enables the projects to become genuinely energy-efficient with minimal environmental impact, fostering a harmonious coexistence between human and the natural environment throughout the landscape architecture process.

课后作业
After-class exercises

———

讨论题
Discussion

进行景观设计时，可持续设计是可有可无的吗？请用 300—500 字来分享你的看法。

Is sustainable design merely a choice when working on a landscape architecture project? Please elaborate on your perspective within 300–500 words.

设计实践
Design practice

根据章节内容，结合实际案例，请用自己的理解制作一个 3 分钟短片报告，解释景观设计是什么，以及它可以解决什么问题。

Based on the chapter topics and specific cases, please create a 3-minute film clip that defines landscape design and the problems it can solve.

推荐读物
Reference reading

1. Fernando-Galiano, L. ed., "Weekend House." *AV Monograph 121 Sanaa: Sejima & Nishizawa*, 2006(8): 112-115.
2. Oh, J. E. and Ma, H. "Enhancing visitor experience of theme park attractions: Focusing on animation and narrative." *Journal of Advanced Research in Dynamical and Control System*, 2018(4): 178-185.
3. LLynch, K. *The Image of the City*. MIT Press, 1964.
4. McHarg, I. L. and American Museum of Natural History. *Design with nature*. American Museum of Natural History, 1969.
8. Starke, B.W. and Simonds, J.O. *Landscape Architecture: A Manual of Environmental Planning and Design*. McGraw-Hill, 2013.
9. Thompson, I. *Landscape Architecture: A Very Short Introduction*. Oxford University Press, 2014.

第二章

自然环境

Chapter 2
Natural environment

学习要点
Study Points
景观设计中自然环境的基本属性和设计原则
The basic properties and design principles of the natural environment in landscape architecture

学习目的
Teaching Objectives
认识自然环境与设计的关系
To understand the relationship between the natural environment and design

本章主要探讨景观设计与自然环境的两个重要关系——气候特征与微气候反应。我们将从气候与人居环境、微气候对设计策略的影响、气候适应性策略、植物种植四个方面展开介绍。重点介绍景观设计如何应对气候环境的变化，形成相应的适应性优化策略。

This chapter delves into two significant relationships: the relationship between landscape architecture and the natural environment, and the correlation between climatic characteristics and microclimatic responses. This analysis will revolve around how landscape architecture adjusts to climate characteristics and the microclimate response in the design of human living environments. Specifically, we will focus on three key aspects: climate characteristics, patterns of climate change, and the social impact of climate on this domain.

一、气候与人居环境

气候对人居环境的影响较大，主要体现在气候特征、气候规律、人文特征三个方面的影响。

1. 气候特征对人居环境的影响

气候是多种变化因素相互作用的结果，这些因素包括温度、水蒸气、风力、太阳辐射、降水等。气候和地形、植被、水体一样，是景观设计中重要的自然环境因素，也是无法避免的设计条件。气候是人类选择栖息空间的主要考量之一。在人居环境的建设中需要关注环境的自然宜居，其中包括寻找舒适及免受自然灾害破坏之地等，自古以来这都是建筑和景观设计主要关注的问题。[29]

随着人类建造环境高速扩张，自然环境被破坏的速度也在不断攀升。其中全球变暖的气候问题已经严重威胁到人类的生存与发展。气候变化过程中所产生的疾病、瘟疫和突发卫生情况也是近几年非常突出的环境问题。[30] 如何应对气候的变化，从而推动人居环境和自然环境的空间共存成为一大问题。而景观设计应该如何适应这一变化？我们需要对区域气候的优势部分进行合理的利用，而不是站在对立面依靠人工建造技术对抗气候变化，这种顺势而为的设计理念与中国造园的"因地制宜"理念不谋而合。这提醒我们在进行设计前，需要对区域的气候条件、温度湿度、气流变化等进行综合分析，从而在基地选址、施工方式、植物配置、建筑设计策略方面最大化利用区域气候的优势，从而达到自然环境资源的可持续调配。[31]

2. 气候的规律变化对人居环境的影响

气候的显著特征包括年度、季节和日间温

2.1 Climate and human settlements
2.1.1 The impact of climatic features on the habitat
Climate emerges as the result of numerous variables, including temperature, water vapor, wind, solar radiation, and rainfall. In landscape architecture, climate assumes an inescapable design condition, holding equal importance to topography, vegetation, and water. Moreover, climate profoundly influences the primary considerations in human habitation selection. As humans coexist and develop alongside the natural environment, it is necessary to attend to the natural habitats when constructing settlements, seeking out comfortable havens that remain unaffected by natural disasters. Throughout history, this concern has remained central to the fields of architecture and landscape architecture.[29]

With the rapid development of the built environment, the rate of destruction inflicted upon the natural environment has surged. Among various factors, climate change has emerged as a grave threat to human survival. In recent years, diseases, epidemics, and health emergencies owing to climate change have assumed prominent crisis proportions.[30] How can we address climate change by creating spaces conducive to harmonious coexistence between humans and the natural environment? Landscape architecture adapts to climate change by harnessing the full potential of regional climates, rather than relying solely on artificial construction techniques to combat it. This design concept aligns with China's notion of "adapting to local conditions" in landscaping. It also serves as a reminder that prior to designing the landscape, it is necessary to comprehensively analyze changes in climate conditions, temperature, humidity, and airflow in the area. This comprehensive analysis enables us to fully consider the climatic advantages in site selection, construction methods, plant configuration, and architectural design strategies, thereby realizing the sustainable design of natural environmental resources.[31]

2.1.2 The influence of regular climate change on human settlements
Annual, seasonal, and daily temperature fluctuations exhibit geographical variations contingent upon their respective positions. These differences in climate are

度变化，这些特征受到诸多因素影响，如经度、纬度、日照强度、植被条件、海湾气流、水体、积雪、沙漠等。[32] 另外，这些特征也会随着时间的变化产生一定的规律。例如，全年的太阳高度变化，是形成不同地区气候的要素之一，对场地的设计规划有着重要影响。季节降水量与气候的关系，可以表现为按露、雨、霜、雪的形式记录的降水量及季节的湿度变化。季节性降水变化带来景观场地对于雨水排洪要求的变化，以及地质结构、土壤类型和土层深度的变化。这些都与建造过程中的施工技术有着密切的关系。温度的变化可以根据地理位置划分出寒带、亚寒带、温带、热带等气候带。通过了解不同区域的温度特征，从而在空间的规划和配置上有所注意。气候特征的变化影响着当地建筑构造的方式，并对区域景观风貌的形成

起到重要的基础作用。

3. 气候的地理人文影响

中国的俗语有称"一方水土，养一方人"，其中"水土"可解释为气候特征。这句俗语也说明了气候与人们的生理健康、心理健康、社会习俗等都有着重要的联系，同时提醒着我们在进行规划设计时需要尊重本土地域的特征。

因此，规划设计的前提是对区域的地理人文、居民行为特征进行分析，理解区域环境中的建筑特点、食物偏好、衣着穿搭、文化习俗、娱乐休闲方式等，这些都是设计调研的重要部分。例如，温带和寒带地区的建筑形态会因气候不同而有不同的表现：在温带地区，为了躲避夏天的高温，大部分公共空间都以风雨连廊的方式进行建造，以增加阴凉的避暑环境；相反，寒带地区为了减少冬季的暖气消耗，建筑

influenced by factors such as longitude, latitude, solar irradiance, vegetation status, hydrological conditions, snowfall, and arid regions.[32] Conversely, all these distinguishing attributes undergo temporal fluctuations in a somewhat regular fashion.

The magnitude and intensity of solar altitude variations throughout the year constitute a pivotal determinant in the climate, thereby shaping the distribution of sunlight across diverse regions. Therefore, consideration of site-specific design planning assumes paramount significance.

The relationship between climate and seasonal precipitation can be documented in the form of seasonal changes in dew, rainfall, frost, or snow, as well as humidity levels corresponding to the different seasons of the year. Regional variations in seasonal precipitation have also necessitated changes in rainwater drainage requirements in landscape sites, alongside changes in geological composition, soil classification, and soil depth. All these factors are closely interlinked with the technological aspects employed during the construction process.

Temperature fluctuations, contingent upon different geographical locations, can be categorized into various climatic zones, including the frigid zone, cold temperate zone, temperate zone, warm temperate zone, and tropical zone. The temperature characteristics exhibited in different regions warrant attention in terms of area planning and configuration. Climate variations have exerted a profound influence on local architectural practices, playing a pivotal role in shaping the landscape of this area.

2.1.3 Climate variations on local architectural practices
A Chinese proverb aptly states, "One side's soil and water nurture one side's people." In this context, "soil and water" can be interpreted as climate characteristics. This adage highlights the significance of climate in relation to physical well-being, mental health, and societal customs. Moreover, it serves as a reminder that due regard for local cultural attributes is necessary throughout the planning and design process. In order to comprehensively address the intricacies of planning and design, a thorough analysis of behavioral patterns must be conducted, taking into account regional climate, geography, and

开窗朝向以日照高度的朝向为主，并减少背光面的开窗面积，从而进行建筑的保温。气候还影响着人们的性格和公共社交方式，如温带和热带区域的居民对户外活动更为向往，总体个性奔放而热烈，更喜欢聚集于广场进行休息和餐饮活动（图2-01）。而对于阴雨天气较多的区域，人们虽然也喜欢在户外，但会选择有连廊的公共空间进行聚会。从区域的气候差异观察不同地理环境所形成的空间和人文变化，以及从文学、音乐和艺术中洞察区域特征和居民特点，都是景观设计必要的调查方式。

二、微气候对设计策略的影响

微气候对于设计策略的影响主要集中在三个方面，分别是规划设计、建筑设计、景观设计。

1. 微气候对规划设计的影响

城市设计如果忽略对气候的考量，那么未来的城市将会产生巨大的能源浪费。近年来，气候危机最主要的问题之一是全球变暖，其中"热岛效应"的产生就是由于城市规划中的微气候调节没有达到较好的平衡。位于城市中心的高楼大厦占据了大部分的用地空间，大量的硬质铺装和玻璃反光材料造成日光的强反射，从而增加了该片区的温度。面对城市发展的需求，如何平衡和调整微气候成为设计师需要考虑的问题。城市设计对微气候进行调整，主要可遵循两个原则。第一，减少硬质铺装的面积。硬质铺装具有强反射性，同等面积的绿地和硬质铺装地面，后者的地面温度可升高3℃。[33]不过，我们也不能过于理想化地把所有地方都变为绿色种植区，屋顶、路面、广场、停车场

the inhabitants themselves. This multifaceted approach allows for a comprehensive understanding of architectural characteristics, culinary preferences, sartorial choices, cultural customs, and modes of entertainment in a given environment. These facets collectively constitute vital components of design research. By examining architectural features, it becomes evident that the architectural forms in temperate and cold regions differ in response to the prevailing climate. For instance, in temperate regions, the construction of public spaces often takes the form of covered walkways, serving as storm corridors that provide respite from the scorching summer heat, thereby affording shade and shelter. Conversely, in colder areas, optimizing heating efficiency during winter necessitates orienting windows towards the sunlight to enhance insulation. Climate exerts a profound influence on the disposition and social inclinations of individuals. For instance, denizens of temperate and tropical regions exhibit a proclivity for outdoor activities, characterized by vivacity and enthusiasm. They tend to congregate in public squares for relaxation and dining (Fig 2-01). In the case of Chongqing, a city renowned for its rainfall in China, locals prefer congregating outdoors, albeit in public spaces equipped with connecting corridors. Analyzing the spatial and cultural differences across regions with varying climates, and gaining insight into regional idiosyncrasies, assumes paramount importance in landscape planning and design research.

2.2 The impact of microclimate on design strategies
2.2.1 The impact of microclimate on urban planning
From the perspective of future urban energy consumption, urban design bereft of climate considerations would be profligate. The prevailing climate crisis, epitomized by global warming, has contributed to the heat island effect, a consequence of the imbalanced microclimate adjustments in urban planning. The construction of impermeable surfaces and glass structures dominates urban cores, resulting in the intense reflection of sunlight and subsequent temperature elevation, thereby giving rise to the aforementioned heat island effect. In light of urban development, designers are confronted with the necessity of reconciling and adapting to microclimate fluctuations in urban design. The adjustment of urban design to

等区域依然需要使用大面积的硬质铺装。因此，根据可持续理念，可将铺装的材料改善为低反射的混凝土材料，从而改善硬质铺装的地面温度。第二，增加阴凉的面积。除了大面积的树木种植外，盆栽植物、廊架藤蔓植物的种植，设置悬挂的遮阳伞等都是增加阴凉面积的好方法。另外，水池、喷泉、喷雾的设置不仅可以降低周边环境的温度，还可以湿润植被。在美国优秀的城市设计案例中，波士顿的"翡翠项链"[34]通过绿地公园的方式植入城市的规划中，从而从根本上控制城市密度的增加（图2-02）。将公园步道植入城市步道中，从而通过植物吸收碳排放以及增加市区的阴凉面积。波士顿"翡翠项链"的规划理念很好地展示了可持续设计的理念在城市发展过程中的重要性。

2. 微气候对建筑设计的影响

建筑设计需要发挥微气候的适应能力，以更好地节约能源。在进行建筑设计的过程中，需要从选址、空间布局、能源消耗等方面来思考微气候的适应性。[35] 在选址方面，建筑的朝向会影响室内的光照需求以及空气对流的质量。例如，在意大利和北非的城镇，建筑需要考虑隔热效果，所以古罗马的很多房屋都没有窗户，加厚的墙壁也使得内部空间较为阴凉。使用内部庭院连廊的空间组合，也可以避免大面积的阳光直射，保持室内空间的阴凉。在中国广东，同样有着对微气候进行调整的建筑形式。广东地区常年炎热，为了避免大面积的地面暴晒和室内温度过高，形成了"骑楼"[36] 的建筑形式。骑楼与古罗马的连廊有相似之处，都是为炎热天气下的户外活动提供遮阴的建

accommodate microclimate variations can be defined by two overarching principles.

Firstly, to mitigate the adverse effects of hard pavement, it is necessary to reduce its extent. Hard pavement, characterized by its high reflectivity, possesses the capacity to elevate ground temperatures by 3 Celsius degrees[33], rendering it a formidable obstacle to the creation of green spaces. Considering the substantial employment of hard pavements in the construction of building roofs, pavements, plazas, and parking lots, a complete transformation of these areas into verdant plantations is not a viable proposition. In sustainable development, the enhancement of paving areas can be achieved through the utilization of low-reflective concrete materials, thereby ameliorating the surface temperature of these rigid surfaces. Second, expanding shaded areas. In addition to large-scale tree planting, several effective strategies can be implemented to increase shade coverage: container gardening, cultivation of trellised climbing plants, and installation of suspended shade sails. Water features such as ponds, fountains and misting systems serve dual purposes—they effectively lower

ambient temperatures while maintaining soil moisture for surrounding vegetation, simultaneously casting cooling shadows on walls and paved surfaces. Under sustainable design principles, microclimate regulation can be achieved through material selection by replacing traditional pavement with low-reflectivity concrete surfaces, coupled with strategic expansion of shaded zones. This integrated approach demonstrates how passive cooling techniques synergistically enhance thermal comfort in urban environments. A notable example of urban design excellence in the United States is the Emerald Necklace[34] in Boston (Fig 2-02), which has been seamlessly integrated into urban planning through the incorporation of green parks, effectively curbing the increase of urban density. As illustrated in Figure X, the park footpath harmoniously coexists with the urban footpath, thereby facilitating the absorption of carbon emissions by the surrounding vegetation and enhancing the prevalence of shade in the cityscape. The Boston Emerald Necklace serves as a compelling testament to the indispensability of sustainable design principles in urban development.

图 2-01　**Italy street cafe**
意大利街头咖啡厅

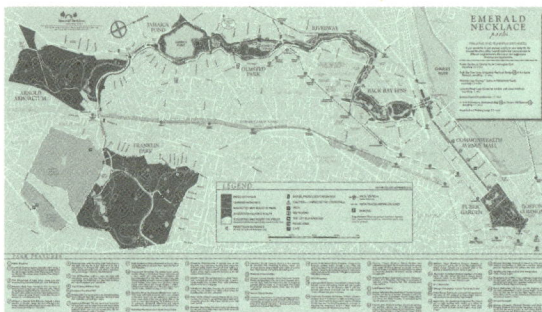

图 2-02　*Emerald Necklace* **in Boston**
波士顿 "翡翠项链"

2.2.2 The impact of microclimate on architectural design

Architectural design must harness the potential of subtle climatic adaptability to optimize energy efficiency. Throughout the architectural design process, consideration must be given to microclimate adaptation, including three fundamental aspects: site selection, spatial layout, and energy consumption.[35] With regard to site selection, the orientation of the building assumes paramount importance as it profoundly impacts the requirements for indoor lighting and the quality of air convection. For instance, in the towns and cities of Italy and North Africa, where thermal insulation is necessary, numerous ancient Roman dwellings were devoid of windows, with their thick walls serving as effective insulators, thereby ensuring a cooler indoor environment. The interior courtyard galleries were ingeniously employed as multifunctional spaces, effectively shielding vast expanses from direct sunlight while concurrently maintaining a cool ambiance in the rooms.

In Guangdong, China, architectural forms have been developed to adapt to the local microclimate. Given the region's perennially hot weather, the architectural concept of the "arcade"[36] emerged as a means to mitigate extensive ground exposure and high indoor temperatures. Resembling the Roman colonnade, the roof structure serves as a shading device, facilitating outdoor activities during hot weather conditions. Moreover, the ramp's height is elevated to 4-5 meters, surpassing the standard building specifications of 3 meters, with the intention of enhancing ventilation. The arcade, functioning not only as a corridor but also as a fusion of shops, passageways, and residences, enables continuous internal air circulation from the corridor to the indoor stores, thereby establishing a natural air-conditioning mechanism. Architects worldwide exhibit boundless creativity in devising strategies to accommodate diverse microclimates. Central European traditional residential forms, for instance, incorporate thick walls to insulate against cold temperatures. Conversely, in hot and humid regions, verandah-like structures with ventilation and shading elements are employed to regulate indoor temperatures. These construction techniques are commonly employed by architects.

筑。骑楼的层高可以达到4—5米，高于一般建筑规范的层高3米，这是为了营造更好的通风效果。与此同时，骑楼的建筑功能并非只是连廊，它还可以集连廊、商铺和居住于一身。较高的楼层设置让空气得到很好的流通，从连廊到室内商铺产生连续的内循环，形成天然的空调模式。在世界各地的不同区域，建筑师面对微气候有无限的设计创意。

3. 微气候对景观设计的影响

景观设计主要通过地形、水、风向、植物、土壤五个方面对微气候进行调节。从我们爬山的体验来说，山上的温度总是比山脚低。夜间峡谷的温度也往往比山坡要低12℃，而湿度要高出20%。同时，清晨山谷容易因为温度的降低而形成雾气，不是修建主干路的理想地点。有一种地形形态称为"霜穴"[37]，这种地形的高处有较多植物阻碍空气的流通，谷底则因为高处树木的遮挡而形成较为暖和的区域。这种地形易在高处山谷形成温度较低的霜地，不适合居住。这种两极分化的坡地情况，说明了室内供暖的成本与户外的环境变化有着密切的关系。因此，住宅的选址最好位于南向坡地的半山腰。作为中国传统民居的一种，甘肃西峰"窑洞"也是利用地形山势而形成的建筑形态。窑洞通过地势与黄土层的堆积而形成内向型的建筑空间。因为黄土土质密实，所以建立的土墙具有很好的挡风保暖效果，为常年受到风沙侵扰的当地居民提供了坚固、耐用的居处。

水具有调节气温的功能，在古代园林和现代城市景观的设计中都具有较好的供暖和制冷作用。水域的规模越大，对微气候的影响就越大。在海洋、大型湖泊和小型池塘附近，都会

2.2.3 The impact of microclimate on landscape architecture

Drawing from mountaineering experiences, it is evident that temperatures at the mountain summit are consistently cooler than those at the base. During the nighttime, the valley temperature often plummets by 12 degrees Celsius compared to the hillside, accompanied by a 20 percent increase in humidity. Additionally, due to the temperature drop, fog frequently envelops the valley in the morning, rendering it unsuitable for constructing main roads. A topographical feature known as a "forest pocket"[37] exists, characterized by denser vegetation on higher ground obstructing air flow, resulting in warmer valleys due to the shade cast by the elevated terrain. This topographical model creates inhospitable mountainous regions, valleys, and cold areas. The polarised sloping landscapes highlights the complex relationship between interior heating costs and outdoor environmental fluctuations. Optimal residential locations are situated halfway up south-facing slopes. Among China's traditional dwellings, the "cave dwelling" in Xifeng, Gansu, exemplifies a settlement that capitalizes on the mountain's topography. The kiln, an inward-looking architectural space formed through the accumulation of yellow soil, leverages the dense loess soil to construct earth walls that effectively shield against sandstorms, which plague Gansu throughout the year.

Water resources possess the capacity to regulate temperature, a phenomenon observed throughout history in various settings ranging from ancient gardens to contemporary urban landscapes. The deliberate design of these spaces aims to optimize heating and cooling effects. Notably, the magnitude of water bodies directly correlates with their influence on microclimate regulation. Proximity to the sea, expansive lakes, and even small ponds can effectively cool ambient air temperatures through the process of surface evaporation, thereby generating refreshing breezes. This explains the moniker "Gold Coast" accorded to Chicago, a city situated in close proximity to Lake Michigan, where summer temperatures can soar to a scorching 30 degrees Celsius. The difference between the lakeside temperature and that of the inland areas, located merely one kilometer away, amounts to nearly 12 degrees Celsius. In the context of classical Chinese gardens, the reflective

因为水面气温较低而形成微风。这也是为什么夏季气温高达 30℃的芝加哥，会因为地理位置靠近密歇根湖而拥有"黄金海岸"的称号。与相距 1 千米的内陆比较，有近 12℃的温差。在中国的古典园林中，人工湖水的水面反射不仅是造园师拓展空间视觉的方法，更是调节室内气温的重要手段。在苏州园林中，多处的连廊和观景楼阁都与湖面相接，利用水面形成的微风使连廊室内空气流通，能够大幅降低室内温度。在现代城市景观中，小喷泉、镜面水池、水幕和水雾气等水景营造都是缓解城市热岛效应、调节微气候的重要手法。

风向对于一个城市的温度有着重要的影响。景观设计可以通过植物和空间形态的营造达到较好的通风效果，并且利用风道形成遮阴的廊道来促进空气的循环，从而调节气温。风向受到地形变化的影响，也会形成不同区域的微气候变化。在迎风坡，气团爬坡而聚集大量的冷空气，从而形成降雨，所以迎风坡通常较为湿润，植被较为密集。相反，在背风坡，气团不断下降而变得干燥暖和，因此降雨较少。[38] 我们可以从中得出利用风向调节城市微气候的方法。例如，在迎风的坡地设置挡墙和树林，可以将空气引导到狭窄的街道空间，从而增加通风的循环。微风经过排列整齐的建筑物、墙体、绿篱、行道树时也会因"狭管效应"[39] 而将风势放大，从而让区域变得更为凉快。

在第一节我们提到通过减少硬质铺装的面积可以降低区域的温度，除此之外，利用植物的生态特点也可以营造可持续的微气候条件。植被如同天然的隔热膜，是建筑空间节约能源消耗的重要手段。可以通过乔木、灌木、地被

properties of artificial lake surfaces serve not only as a means for gardeners to expand their spatial perception but also as a crucial element for regulating indoor temperatures. In Suzhou gardens, for instance, interconnected corridors and viewing pavilions are strategically positioned alongside lakes, facilitating the circulation of air through the generation of breezes, thereby significantly reducing indoor temperatures. In modern urban landscapes, the incorporation of small fountains, mirror pools, water curtains, water mist, and other water features assumes paramount importance as they effectively mitigate the urban heat island effect and contribute to microclimate adjustment.

The direction of prevailing winds assumes a pivotal role in influencing urban temperatures. By employing strategic landscaping techniques, such as the creation of green spaces and the establishment of spatial configurations, enhanced ventilation can be achieved, thereby fostering the formation of sunshade corridors that facilitate air circulation and temperature regulation. When air currents are propelled by prevailing winds, the upward movement of air masses results in the accumulation of copious amounts of cold air, and finally, precipitation. Therefore, windward slopes tend to be characterized by higher levels of humidity and denser vegetation. Conversely, leeward slopes receive minimal rainfall, causing descending air masses to become dry and warm as they continue their descent[38]. The topography of a given region contributes to variations in wind direction, thereby resulting in microclimate temperature fluctuations across different areas. From this, we can deduce a methodology for adjusting urban microclimates by harnessing wind direction. For instance, the strategic placement of retaining walls and trees on windward slopes can effectively channel wind currents into narrower street spaces, thereby enhancing ventilation. As the breeze traverses neatly arranged buildings, walls, hedges, and street trees, it undergoes amplification through the phenomenon known as "the effect of narrow,"[39] thus creating a cooling effect in the vicinity.

Regarding planting, we have previously highlighted the role of reducing impermeable road surfaces in lowering temperatures in a given area. By capitalizing on the ecological attributes of plants, it becomes possible to

的不同组合，形成高低错落的空间形态，借助植物疏密的设计来达成不同的通风遮阴效果。例如，稀疏的阵列树木，能引导夏季的风向，从而形成连续的微风效应[40]，而密实的灌木排布可以阻挡冬季寒风并遮挡夏季的炎热。

土壤的变化也可以小幅度地影响气温。干燥的沙土、沙砾温度较高而湿度较低，容易在炎热的天气下影响地面的温度。湿润的土壤包括排水性较差的沼泽黏土，往往温度低，而湿度较高。这些土壤的特质能在特定的区域帮助建筑师进行选址，从而设计出更为节能的建筑。

三、气候适应性设计策略

人类需要与自然环境共存发展，人居环境需要面对气候进行适应性的调整。根据城市设计的需求，我们将地球分为四个基本气候带：

寒带、亚寒带、温带、热带（沙漠地区）。每一个气候带都有着在气候条件影响下形成的独特建筑风貌。虽然极寒和极热地区少有人栖息，但建筑师也需要对该地区极端天气所带来的问题进行归纳总结，才能在城市规划和设计的过程中及早预防和解决。与此同时，我们也可以在不同的气候条件下，观察设计所做的策略调整，从而在面对不同的条件和气候环境时，都能为场地建立适应性较好的社区布局、场地规划和景观设计的方案。

1. 寒带

寒带位于地球的高纬地区，在极圈以内，气候特征以极寒天气为主，终年持续低温，所以称为寒带。极寒气候的优势在于对自然风光的保护，集中了世界知名、别具特色的极光观赏地，如雷克雅未克（冰岛）、基律纳（瑞典）、

establish sustainable microclimate conditions. The improvement of vegetation assumes a pivotal role in natural thermal insulation, thereby constituting a crucial means of conserving energy. Through the strategic combination of trees, shrubs, and ground-cover plants, a highly staggered pattern can be achieved, thereby facilitating ventilation and shading effects through plant design. For instance, a sparsely arranged array of trees can effectively guide wind currents during summer, thereby creating a continuous breeze.[40] Conversely, densely packed shrubbery serves to impede cold winds during winter and mitigate heat during summer.

Changes in the soil can exert a modest influence on temperature. In hot weather, dry and sandy soil, characterized by higher temperatures and lower humidity, impacts ground temperature. Conversely, moist soils, particularly poorly drained swamp clay, tend to exhibit lower temperatures and higher humidity. This change in soil temperature serves as a valuable criterion for site selection in specific regions, facilitating the design of more energy-efficient building models.

2.3 Design strategies for climate adaptation

Considering the necessity for harmonious coexistence and development between humans and the natural environment, it becomes necessary to adapt to prevailing climatic conditions. Broadly speaking, the Earth can be categorized into four climatic zones: the cold zone, sub-frigid zone, warm temperate zone, and dry tropical zone. Each climatic zone possesses its own architectural landscape, which is profoundly influenced by climatic factors. While sparsely populated, regions experiencing extreme cold and hot weather warrant consideration of design challenges in order to preempt and address these issues during the urban planning and design process. Concurrently, it is crucial to assess how design strategies adapt to diverse climatic conditions, thereby creating better solutions for sites situated in dissimilar circumstances and climates, thus facilitating the harmonious integration of community layout, site planning, and landscape architecture.

2.3.1 Cold zone

The cold zone is located in the high-altitude area of the

罗瓦涅米（芬兰）、圣诞小镇德勒巴克（挪威）等。极寒气候下需要应对的问题包括交通、积雪防滑、建筑的保温抗灾能力等。极寒地区在规划城市设计的时候，需要充分考虑以下设计要点：

（1）将居住区域和文娱、商业、医疗中心集中布局，减少路上交通时间。

（2）限制规划尺度以减少高昂的开发和防霜冻建构。

（3）建筑朝向温暖阳光，尽可能利用日照，以达到充足的自然光照。

（4）建筑采用前倾屋顶，加深屋檐，增强风暴雨水天气下的排洪的能力。

（5）将道路安排在阴影带内以防止结冰。

2. 亚寒带

亚寒带又称副北极带和副南极带，冬季盛行极地温度，夏季盛行温带温度。气候特征全年温差较大，冬季严寒而漫长，夏季温暖而短暂。降雨量少，部分位于亚寒带的地区还会伴随霜降和降雪的出现。气候植被主要位于欧亚大陆和北美大陆北部，以针叶林为主。亚寒带气候多变，也具有部分强降雨、水分充足、地貌多河流和淡水的地区。针对冬季寒冷而日照时间短的气候条件，城市规划设计需要考虑出行便利和节能保温的方式。其中具体要点如下：

（1）调整街道和开放空间的形态引导冬季风的冲击，并增强夏季微风。

（2）街道市政的排洪排水功能需要满足极端天气下的需求。

（3）公共空间需要具备灵活可变的形态，以满足春夏、秋冬不同的户外活动需求。

（4）建筑设计要最大限度地利用气候特

Earth, within the polar circle, and it is extremely cold all year round. It is named the northern belt. The advantage of the cold climate is the preservation of natural beauty, and the landscape of the world's most famous cities for viewing the Aurora Borealis is distinctive, including Reykjavik (Iceland), Kiruna (Sweden), Rovaniemi (Finland), and Drebak (Norway). In the face of extremely cold weather, the region has to deal with traffic problems, snow, and anti-skid problems, as well as the insulation and elasticity of buildings. The following design points need to be considered when planning urban design in extremely cold areas:

▪ Centralize the layout of residential areas and other cities, businesses, and medical centers, to reduce travel time.
▪ Limit planning scales to reduce high-cost development and frost-proof construction.
▪ Orientate buildings towards sunlight, making use of daylight wherever possible to achieve sufficient natural light.
▪ Build buildings with a forward-sloping roof, deepening the eaves and the ability to drain rainwater from large storms.
▪ Position roads in shaded areas to prevent ice from forming.

2.3.2 Sub-frigid zone

The sub-frigid zone is situated in the sub-Arctic and sub-Antarctic regions, where polar temperatures dominate during winter and mild air masses prevail in summer. This climate exhibits significant temperature fluctuations throughout the year, characterized by long, cold winters and short, warm summers. Precipitation is scarce, and certain areas lie in cold temperate regions, experiencing frost and snow. Climatic vegetation is predominantly found in Eurasia and the northern part of North America, with coniferous forests being the most prevalent. The variability of the cold temperate climate is accompanied by significant rainfall, abundant water resources, and numerous rivers and freshwater landforms. Considering the challenges posed by cold winters and limited sunlight, urban planning and design must consider convenient transportation, energy conservation, and thermal insulation. The specific details are as follows:

▪ Transform the layout of streets and open spaces to guide the influence of winter winds and enhance summer breezes.
▪ The municipal administration responsible for street management should ensure that flood control and drainage

征满足制冷和通风的要求。

（5）结构设计需要能够应对收缩、膨胀、凝结、冰冻等四季温差导致的变化。

（6）通过本地植物和天然自然景观增加区域景观特色。

（7）亚寒带典型地区案例：北欧地区（图2-03）、加拿大（图2-04）、中国哈尔滨（图2-05）。

3. 温带

温带是位于亚热带和亚寒带之间的气候带。我国大部分地区处于温带。温带气候特征为四季变化分明，温差较大，是十分适合居住的气候条件。四季分明的气候带来丰富的植物形态，如落叶阔叶树种和针叶树种混合的森林形态。温带景观地貌丰富多样，有海滨、平原、高原、山地等。对于温带的城市景观设计，需要考虑到四季变化的户外空间休闲需求，并将强降雨、炎热高温、大风、洪水等气候问题的应对方式纳入公共空间设计。在景观规划策略中，需要充分利用多彩丰富的植物景观营造独特的自然风光，并注意降温通风。其中具体要点如下：

（1）结合树荫和水资源为空间提供遮阴、通风的效果。

（2）城市排水系统需要具备应对特大暴雨的能力。

（3）通过合理规划树木配置，减少强光对地面的照射。

（4）空间规划需要考虑早晚活动的分流，避免公共空间在白天无遮阴场地。

（5）可考虑抬高建筑并设置地下通风装置，以减少蚊虫和潮湿的侵扰。

systems can withstand extreme weather conditions.
- Public spaces should adopt flexible and adaptable approaches to accommodate seasonal changes and meet the demands of year-round outdoor activities.
- Buildings should leverage the climate's characteristics and fulfill cooling and ventilation requirements.
- Structural design should be capable of adapting to seasonal temperature fluctuations, including contraction, expansion, condensation, and freezing.
- Enhance the characteristics of regional landscapes through local plants and natural scenery.
- Typical Urban Cases in Cold and Temperate Zones: Northern Europe (Fig 2-03), Canada (Fig 2-04), Harbin, China (Fig 2-05).

2.3.3 Warm temperate zone

The warm temperate zone is situated in the subtropical zone and extends up to the polar circle. Most regions in this zone enjoy a mild climate. The temperate climate is characterized by distinct seasons and significant temperature variations. Winter in the temperate zone is cold, while summer is hot, making it an ideal living climate. This zone covers approximately 50 percent of the Earth's total area. The unique climatic conditions have fostered a rich and diverse array of plant forms, resulting in a blend of deciduous broad-leaved forests and coniferous forests. The warm temperate zone features varied and abundant surface vegetation, including coastal, plain, plateau, and mountain landscapes. Urban landscape architecture in warm temperate regions must account for the ever-changing demand for outdoor leisure spaces throughout the year. The design of public spaces should address climate-related crises, such as rainstorms, intense heat, strong winds, and floods. In landscape planning strategies, it is essential to leverage the vibrant and diverse plant landscapes to create unique natural scenery. The key elements of cooling and ventilation design are as follows:
- adequate shading and ventilation for the space.
- Urban drainage systems must be designed to accommodate the drainage capacity required during torrential rain.
- Rational tree configuration planning should reduce or eliminate excessive ground illumination caused by intense lighting.

图 2-03　**Northern European city in Denmark**
　　　　北欧丹麦某城市

图 2-04　**Waterfront planning in Canada**
　　　　加拿大滨水规划

图 2-05　**Old Harbin block**
　　　　老哈尔滨街区

（6）温带典型城市案例：日本大崎（图2-06）、中国北京（图2-07）。

4. 热带（沙漠地区）

热带（沙漠地区）位于赤道线附近，常年炎热、干燥，降雨量少，适合耐旱植物的生长，景观地貌以沙丘和戈壁为主。气候特征包括强光照、空气湿度小、早晚温差极大等。热带沙漠地区常年处于干旱炎热环境，造成大片无植被地区，植物则以稀疏的旱生灌木和少数草本植物，以及一些雨后生长的短生植物为主。热带沙漠气候主要分布在非洲北部、阿拉伯、非洲西南沿海等。在干热沙漠地带，需要重点应对的问题包括极热降温、水资源灌溉、植物的种植和维护等。主要设计要点如下：

（1）紧凑的连廊和串联的庭院，避免太阳对步行区域的照射，同时增加微风对流。

（2）使用本土植物，保护沙漠自然生物。

（3）避免户外公共空间完全暴露。

（4）限制公园、植物种植区的面积。

（5）建立具有废水储水功能的花园灌溉系统，关注水景的配置。

（6）热带沙漠地区典型城市案例：撒哈拉沙漠阿推夫城（图2-08）、澳大利亚蒂布巴拉（图2-09）。

四、景观种植

植物受气候环境的变化而形成丰富的生态。面对不同环境的景观种植，需要先了解植物种植的目的、配置与作用、可持续种植等概念。

1. 植物种植的目的

历史上，第一个意为"景观"的词是用来

• Spatial planning should consider morning and evening activities to ensure that shaded areas are available for public use during the day.
• Buildings can be elevated and equipped with underground ventilation systems to mitigate mosquito infestation and moisture-related issues.
• Typical Urban Cases in Warm and Temperate Zones: Osaki, Japan (Fig 2-06), Beijing, China. (Fig 2-07)

2.3.4 Dry tropical (desert) zone

The tropical desert region is situated in close proximity to the equatorial line, exhibiting perennially hot and arid temperatures alongside scant precipitation, rendering it conducive to the proliferation of drought-resistant flora. The landscape is predominantly characterized by vast expanses of dunes and the Gobi Desert. The climatic conditions prevailing in this location are typified by intense solar radiation, reduced atmospheric moisture, scorching diurnal heat, and frigid nocturnal temperatures. The indigenous vegetation in a tropical desert climate endures a constantly parched and torrid environment, thereby contributing to a lack of greenery and extensive tracts of barrenness. The botanical composition is predominantly comprised of meager, desiccated shrubs, interspersed with a smattering of herbaceous species, as well as a few diminutive plants that sprout following precipitation events. Tropical desert climates are primarily encountered in northern Africa, Arabia, and the southwestern coast of Africa. In arid and torrid desert regions, the challenges of extreme cooling, water resource irrigation, and horticultural maintenance necessitate resolution. The principal design considerations include the following:

• The strategic arrangement of porches and courtyards in a compact, sequential configuration serves to shield the pedestrian zone from direct solar irradiation, while concurrently fostering the convection of breezes.
• Embrace the utilization of indigenous flora, thereby protecting the natural ecosystem in the desert.
• Avoid the complete exposure of outdoor communal spaces.
• Imposition of limitations on the dimensions of parks, ornamental grasslands, and planting areas.
• Incorporation of garden irrigation systems, featuring provisions for wastewater storage and the configuration

图 2-06　**Urban planning of Osaki, Japan**
日本大崎城市规划

图 2-07　**Beijing Olympic Forest Park**
北京奥林匹克森林公园

图 2-08 **El Atteuf in Sahara Desert**
撒哈拉沙漠阿推夫城

图 2-09 **Tibooburra, Australia**
澳大利亚蒂布巴拉

of waterscapes.

▪ Typical case of dry-tropical cities: Sahara desert (Fig 2-08), Australian desert landscape（Fig 2-09）.

2.4 Landscape planting
2.4.1 Purpose of planting

The earliest term denoting "landscape" was employed to denote the modest vegetable garden situated at the rear of a dwelling. The cultivation of plants represents a pivotal undertaking throughout the annals of human history. The propagation and nurturing of plants have been accompanied by diverse characteristics across various periods. During the early stages of agrarian societies, individuals commenced the cultivation of vegetables, grains, fruits, and poultry in enclosures, thereby establishing a stable environment to supplement the sustenance requirements. In ancient Egypt, a significant proportion of the populace resided along riverbanks, thereby benefiting from the ready availability of natural water resources for crop irrigation. Therefore, the productivity of expansive tracts of farmland and vineyards witnessed a marked upswing. Concurrently, humans commenced the construction of barns, subterranean storage chambers, and granaries to ensure the continuity of the food supply. Simultaneously, the natural environment of botanical gardens provided families with spaces for study and repose. During medieval Europe, the exigencies of warfare compelled populations to live in fortified strongholds. The sole verdant expanse in the confines of the courtyard was allocated for subsistence crop cultivation and children's recreational activities. Estate proprietors were compelled to acquire knowledge pertaining to the growth patterns of crops, flowers, and herbs, in order to cater to the diverse dietary requirements dictated by life's exigencies.

The positioning of contemporary urban centers is frequently predicated upon the proximity to rivers and harbors, followed by a complex network of intersecting expressways, thereby establishing the fundamental framework of the city. The evolution of urban areas is intrinsically linked to land exploitation, deforestation, and river contamination, which exert a profound influence on the natural environment. In the present era, planting design

表达房子后面的小菜园的。在人类历史发展的过程中，植物种植始终是人们定居生活的重要活动。植物的繁殖和培育伴随着不同时代背景的特色。在早期的原始农庄，人们开始把蔬菜、谷物、水果、动物等围合种植和圈养起来，形成稳定的食物供给。在早期的埃及地区，大部分居民都沿着河流居住，天然的水资源可便利地为植物和农作物提供灌溉。随着农田、植物园的面积和产量不断增加，人类开始建设谷仓、地下储藏室和饲料储藏室来维持食物的供应。与此同时，植物园的自然环境也赋予家庭学习和休息的生活场所。在中世纪的欧洲，战乱的生活环境迫使人们打造庄园和城堡，仅有的庭院绿地除了种植生活农作物以外，也是妇女和儿童游玩的地方。庄园主还需要学习农作物、花卉、草药等的基本生长习性，以应对不同的生活需求。

综观现代城市的选址，大多以河流和港湾为主要的聚集地，然后规划以纵横交错的高速道路，形成城市的基本架构。同时，城市的发展所伴随的土地的开发、森林的砍伐、河流的污染也大面积侵害了自然环境。在当代，种植设计不仅要满足城市美观和休闲生活的需求，更要注重植物种植对于生态环境的改善，以及是否具有污染过滤和净化的作用。看重可持续生态节能作用的植物种植设计也越来越多，例如，雨水花园[41]要求植物配置具有过滤雨水的能力；低维护的种植方式更多地应用在郊区大面积的道路系统中，有利于雨、废水的收集和自然过滤。

植物种植在现代景观设计中已经逐渐成为节能节水的重要自然手段。植物种植的设

comprises not only urban aesthetics and the fulfillment of leisurely pursuits, but more crucially, it necessitates a conscientious focus on the ecological amelioration of vegetation, as well as its capacity to effectively filter and purify pollutants. Increasingly, planting design seeks to produce a sustainable ecological environment and facilitate energy conservation. For instance, the design of rain gardens mandates the incorporation of rain-filtering flora. In addition, low-maintenance planting is optimally employed in expansive suburban road systems, thereby fostering rainwater collection and natural filtration.[41]

In the domain of contemporary landscape architecture, planting has emerged as a pivotal and innate mechanism for conserving energy and water resources. The design philosophy and conservation practices pertaining to planting also serve as a testament to the symbiotic equilibrium between urbanity and the natural world. The better countries seek to protect their natural environs, the more bountiful the qualitative aspects of the natural landscape and ecological matrix shall become.

2.4.2 Configuration and role of plants

Landscape architecture necessitates a greater understanding of the natural environment, inclusive of flora. Plants grow in a variety of forms, ranging from the majestic redwood forests of the Pacific Ocean to the resplendent money pines of the Yangtze River valley, and from the verdant ferns to the subaqueous algae of the Brazilian rainforest, each possessing unique attributes. Concurrently, plant species in nature serve as a vital component in the biological food chain. The arrangement and selection of plants not only epitomize the climatic characteristics endemic to a given place, but also constitute an integral facet of the local ecosystem. Thus, the reasonable, standardized, and ecologically balanced allocation and selection of plants assume paramount significance in the advancement of regional ecology and urban landscape spaces. Three key areas warrant further attention prior to the configuration of plant species.

Plant identification

The global diversity of plant species necessitates the adoption of an international lexicon to facilitate the description and identification thereof, thereby enabling

计理念和养护规范体现着该地区城市和自然的共生平衡关系。植物森林资源保护得越好的国家，拥有的自然景观品质和生态基质越是丰厚。

2. 植物的配置与作用

景观设计需要更为敏锐的自然环境认知，其中便包括对自然植物的认知。从太平洋的红杉树林、长江流域的金钱松，到巴西热带雨林的蕨类和海底藻类植物，植物的种类和生长形式多种多样。并且，自然界的植物也是生物链中的食物来源。植物的配置与选择，不仅与区域特征的气候形态相关，也是当地生态系统的重要组成元素。合理、规范、具有生态平衡力的植物配置与选择，对于区域生态和城市景观空间发展具有重要意义。

植物配置需要关注以下三个方面。

（1）植物鉴别

全球植物种类多样，需要有国际通用的语言来进行描述和鉴别，以便不同国家的学者研究自然的奥秘。植物命名通常采用科学的双名法（拉丁语）：属名＋种加词。地区的俗名差异会影响对植物的描述，以拉丁文为标准化的植物命名方式为全球进行植物研究和设计带来便利。

从图中可以看到同样品种的植物在拉丁名属性上是一致的，而在英文翻译中各有不同（图2-10）。

（2）植物环境特性

植物的生长需要依赖环境，反过来，植物对环境也具有重要的调节作用，甚至可以改善气候。在冬季，落叶和种植的地被可以保护土壤、抵御寒风和减少水土流失。在夏季，植物

scholars to conveniently investigate the enigmatic facets of nature. Plants are classified and denoted by scientific binomial nomenclature (in Latin), including genus and species. The attributes and classification of plants can be influenced by regional differences, which in turn impact the vernacular designations ascribed to plants. The standardized practice of employing Latin nomenclature has led to a promotion of plant design and research across the globe.

As illustrated in Figure 2-10, the Latin nomenclature attributes of identical plants remain consistent, while the English translations of local descriptions diverge.

Environmental characteristics of plants
The growth of plants is contingent upon environmental conditions, while concurrently assuming a pivotal role in the regulation of said environment. Plants possess the capacity to regulate and ameliorate the climate. During the winter season, the descent of foliage and the cultivation of ground cover serve to protect the soil against the onslaught of frigid winds, thereby averting

soil erosion. In the summer months, plants engage in photosynthesis, thereby replenishing atmospheric nutrients, and evaporate water through their leaves, thereby functioning as an efficacious natural regulator, fostering moisture retention amongst neighboring flora and mitigating ambient temperatures. Plants undergo a life cycle characterized by growth, maturation, decay, and reproduction. In the case of arid plants, decomposing fibers and cells are assimilated from the soil surface, subsequently undergoing transformation into nutrients, thereby enriching the soil. Simultaneously, fruits, decaying wood, stems, and vines that are unable to degrade in the soil are carried away by water currents, thus serving as sustenance for aquatic plants, shellfish, and fish. The life cycle of plants serves to preserve the equilibrium of the natural ecosystem.

Plant aesthetics
From the earliest stages of landscape development, which were primarily centered around subsistence-oriented vegetable gardens[42], to a period wherein individuals began to systematically exhibit their manipulation of nature

Trees & Shrubs

Acer triflorum	three-flowered maple	*Indigofera amblyantha*	pink-flowered indigo
Aesculus parviflora	bottlebrush buckeye	*Indigofera heterantha*	Himalayan indigo
Amelanchier arborea	common serviceberry	*Juniperus virginiana* 'Corcorcor'	Emerald Sentinel® eastern red cedar
Amelanchier laevis	Allegheny serviceberry	**Emerald Sentinel ™**	
Amorpha canescens	leadplant	*Lespedeza thunbergii* 'Gibraltar'	Gibraltar bushclover
Amorpha fruticosa	desert false indigo	*Magnolia macrophylla*	bigleaf magnolia
Aronia melanocarpa 'Viking'	Viking black chokeberry	*Magnolia tripetala*	umbrella tree
Betula nigra	river birch	*Magnolia virginiana* var. *australis*	Green Shadow sweetbay magnolia
Betula populifolia	grey birch	'Green Shadow'	
Betula populifolia 'Whitespire'	Whitespire grey birch	*Mahonia x media* 'Winter Sun'	Winter Sun mahonia
Callicarpa dichotoma	beautyberry	*Malus domestica* 'Golden Russet'	Golden Russet apple
Calycanthus floridus	sweetshrub	*Malus floribunda*	crabapple
Calycanthus floridus 'Michael Lindsey'	Michael Lindsey sweetshrub	*Nyssa sylvatica*	black gum
Carpinus betulus 'Fastigiata'	upright European hornbeam	*Nyssa sylvatica* 'Wildfire'	Wildfire black gum
Carpinus caroliniana	American hornbeam	*Philadelphus* 'Natchez'	Natchez sweet mock orange
Cercis canadensis	eastern redbud	*Populus tremuloides*	quaking aspen
Cercis canadensis 'Ace of Hearts'	Ace of Hearts redbud	*Prunus virginiana*	chokecherry
Cercis canadensis 'Appalachian Red'	Appalachian Red redbud	*Ptelea trifoliata*	hoptree
Cercis canadensis 'Forest Pansy'	Forest Pansy redbud	*Quercus macrocarpa*	bur oak
Cercis canadensis 'Pauline Lily'	Pauline Lily redbud	*Rhododendron atlanticum*	dwarf azalea
Chaenomeles speciosa 'Toyo-Nishiki'	Toyo-Nishiki flowering quince	*Rhododendron viscosum*	swamp azalea
Chaenomeles x superba 'Jet Trail'	Jet Trail flowering quince	*Rhus aromatica* 'Gro-low'	Gro-low aromatic sumac
Chimonanthus praecox	fragrant wintersweet	*Rhus copallinum*	winged sumac
Chionanthus retusus	Chinese fringetree	*Rhus glabra*	smooth sumac
Chionanthus virginicus	fringetree	*Rhus typhina*	staghorn sumac
Clerodendrum trichotomum	harlequin glorybower	*Rhus typhina* 'Dissecta'	cutleaf staghorn sumac
Clethra barbinervis	Japanese clethra	*Rosa* 'Ausorts'	Mortimer Sackler® rose
Clethra alnifolia	summersweet	*Rosa* 'F.J. Grootendorst'	F.J. Grootendorst rose
Cornus florida	Jean's Appalachian Snow dogwood	*Rosa* 'Sally Holmes'	Sally Holmes rose
'Jean's Appalachian Snow'		*Rosa glauca*	red leaf rose
Cornus sanguinea 'Midwinter Fire'	Midwinter Fire bloodtwig dogwood	*Rosa virginiana*	Virginia rose
Cornus 'Rutban' AURORA	Aurora® dogwood	*Rubus rolfei*	creeping raspberry

图 2-10 **List of plant species**
植物种类拉丁文标注名录

可通过光合作用从空气中吸收养分，叶片蒸发的水蒸气有利于保持周边植物的湿润，还能自然调节区域温度。植物具有全生命周期的生长、老化、降解、再生产的能力。对于干枯的植物，腐烂的纤维和细胞被土壤表层吸收，重新转换成养分，从而形成土壤新的生产力。不能在土壤中完全降解的果实、腐木、茎藤会被冲刷进海洋，成为水生植物、贝类和鱼类的养料。植物形成的全生命周期的自然生态系统，一直在维护着自然生态的平衡。

（3）植物美学

从最开始以食物供给为目的的蔬菜园[42]，到人们通过经验和科学的认知开始有计划地对自然进行塑造，园艺美学成为展示国家或区域人文特色的重要视角。在王权至上，对图案和几何形式痴迷的法国古典主义时期，植被被大量用以营造园艺中的地毯图案式样。意大利文艺复兴时期的科学和人文发展，使意大利园林种植引入了精湛丰富的植物花期变化以及理水造景的研究，营造出更诗意和梦幻的情境。东方园林以中国和日本为典型。日本国土面积虽小但地形变化丰富，又四面环海。园林中以植物和园艺表达山与海的景观之美，形成了枯山水[43]、龙华寺园林等古典精致之美。中国地广物博，国土拥有丰富的景观特征，植物种植更是讲究人文寓意。在苏州耦园的入口，迎客松作为主景的植物形态，与墙壁的诗词呼应形成主人对于精神世界的理解。植物的美学在不同国家和地区都有着无限变化的可能，这也是它们经历了无数世纪后留存下来的自然力量的结果。

through experiential and scientific knowledge, horticultural aesthetics assumes a paramount role in explaining the human character of a country's regions. In France, imperial authority reigns supreme, with pattern and geometric forms prevailing. The classical era witnessed extensive utilization of vegetation to create carpet-pattern style horticulture. The scientific and humanistic splendor of the Italian Renaissance endowed Italian gardens with an opulent array of planting variations and an exploration of water management techniques, thereby conferring upon Italian gardens a more poetic and ethereal ambiance. Japan, being a small country with complex and diverse topography, reflects the beauty of mountainous and coastal landscapes in its gardens, thereby producing classical and exquisite beauty, exemplified by Japanese dry gardens[43] and the Longhua Temple Garden. China, featuring an expansive territory replete with diverse landforms, resulting in plant displays that embody the symbolic significance of humanistic ideals. At the entrance to the Suzhou Ou Yuan Garden, the venerable pine tree, in conjunction with the poetic verses adorning the walls, serves as a testament to the proprietor's profound understanding of the spiritual realm. The aesthetics of plants undergo perpetual transformation across different countries and regions, a reflection of the forces of nature that have persevered throughout the centuries.

2.4.3 Sustainable landscape planting principles Philosophy

Planting is not solely focused on urban open spaces or serving as a complement to buildings; rather, it plays a pivotal role in shaping the regional culture of a place. The act of landscape planting serves as a means for designers to convey the ambiance of a location. By virtue of the cultural attributes and environmental considerations embodied by plants, they contribute to the preservation of natural environments in cities, including mountains, rivers, forest wetlands, and oases. Planting aids in maintaining the equilibrium of soil and water, regulating the climate, mitigating wind and sand, and fostering the creation of diverse and vibrant landscapes. As nature and habitats evolve, landscape planting enables the establishment of green and ecological natural areas, thereby facilitating living in communities enveloped by natural open spaces.

3. 可持续理念下的景观种植原则

（1）理念

植物景观不仅是城市空地的点缀或建筑的陪衬，也往往奠定着场地地域文化的基础，是场所气氛的传达者。植物传递的文化品格和爱护环境的提示，有助于促进城市对于自然山河、森林湿地、沙丘绿洲等环境的保护。植物有助于保持水土平衡、调节气候、防御风沙，而且能辅助营造丰富的地形景观。景观种植可以帮助我们创建绿色生态的人居环境，让我们生活的社区被自然开放的空间包围，让我们在日常休闲和公共绿地中感受自然的丰富美感，随着季节的变化观赏花园、植物园、果园以及森林公园带来的缤纷世界。

（2）景观种植导则

景观种植需要考虑场地条件、植物层次的作用、本土植物的选择三个方面。景观种植的主要宗旨在于：尽可能保护原有场地的种植，尽可能以低建造的方式进行植物设计。在场地条件的调研中，需要以现场植物的点位进行人行步道、车行道路、街道和建筑物的协调调配。尊重原有场地植物的生长环境，特别是对特大乔木和珍稀树种来说，挪动位置容易对树木的根部造成破坏，也会影响树木的健康程度。设计师应根据项目需求，考虑现场植物生态的综合平衡，调整设计方案以最好地适应场地的自然环境。植物的类型分为乔木、灌木、植被三个大类，同时也是植物空间营造的基本元素。乔木一般高2—3米，树木修长而有宽阔的树冠，多用于城市行道树的种植，从而形成连续的林荫空间。乔木的群植形态，也可以为场地提供基础的场景色调。如办公楼前的银杏树阵

Simultaneously, it affords us the opportunity to immerse ourselves in the resplendent beauty of nature through daily recreational and public green spaces, where we can behold the kaleidoscopic world of gardens, botanical gardens, orchards, and forest parks, each metamorphosing with the changing seasons.

Landscape planting guidelines

The planting of landscapes necessitates the consideration of three key aspects: site conditions, the role of plant layers, and the selection of indigenous flora. The primary objective of landscape planting is to preserve the original site planting to the greatest extent possible, while adopting a less contrived approach to the design. The study of site conditions involves a comprehensive assessment of footpaths, carriageways, streets, and buildings, prior to contemplating the placement of plants. Due regard must be given to the growth environment of existing plants, particularly oversized trees and rare species, as the relocation of trees can readily inflict damage upon their roots and impede their healthy growth. In accordance with project requirements, designers should strive to achieve a harmonious balance between the on-site plant environment and ecology, thereby ensuring that the design is optimally suited to the natural surroundings. Secondly, the planting should incorporate different layers. Plant species can be categorized into trees, shrubs, and ground cover. Trees, typically measuring 2-3 meters in height, feature elongated and slender trunks, crowned with expansive canopies, making them well-suited for urban streets, where they provide shade. The dense configuration of trees also offers an aesthetically pleasing visual effect. For instance, the uniform arrangement of ginkgo trees in front of the office building (Fig 2-11) imparts a striking feature to the square. Shrubs, ranging in height from 0.5 to 2 meters, serve as an effective barrier against both visual intrusion and wind. In larger spaces, it is advisable to allocate areas of varying sizes for activities. Mid-level trees can be combined with shrubs to form a natural retaining wall or a plant wall, thereby accentuating the prominence of the main landscape tree (Fig 2-12). Ground cover plants, situated at ground level, facilitate defining the theme of roads and sites. Vegetation, conducive to large-scale planting, allows for the selection of multiple plant types.

（图 2-11），整齐统一的植物样式能够为广场提供引人注目的空间特征。灌木高度约0.5—2米，是较好的屏障，既可遮挡视线又可起到挡风的效果。对于场地较为宽阔的空间，需要做大小不一的活动场所分割，中层的树木可结合灌木形成天然的挡墙，或者背景基调的植物墙，更好地显示主景树的位置（图 2-12）。底层地面种植地被植物，有利于界定道路和奠定场地主题色调。可以采取大面积的种植，选取一种或三到五种植被进行组合搭配。现代花境的植被设计模拟自然环境的野生生境，甚至选取三十到四十种植物进行搭配，以营造不同色彩的季节景观（图 2-13）。根据场地功能的需求，可以对植物采取高、中、低三层空间的组合方式，形成符合场地功能特点的植物设计。

（3）景观种植的可持续设计方法

植物作为天然的可持续资源，把植物应用到城市空间中以调节气候、湿度和温度，是重要的可持续设计方法。其中，森林公园、湿地公园、雨水花园、屋顶花园、绿色景墙，以及自然生境都是常用的形式。在城市规划对区域绿地和生态系统的调配中，森林公园和湿地公园作为面积较大的绿地种植是有助于调节区域气候的重要方法。广阔的绿植空间能容纳上千个植物品种，保护珍稀树种，并形成稳定的生物生态群落，对区域的生态循环系统有着重要的意义。

随着存量时代的发展，城市的绿地面积变得尤为珍贵，单一的绿地休闲空间已经无法满足现代城市空间的多样式发展。不同形式的景观种植不断地被试验，以景观绿植在不同城市

Modern flower borders seek to emulate natural habitats, with the possibility of incorporating 30 to 40 plants to create seasonal groupings exhibiting a diverse range of colors (Fig 2-13). The three plant layers can be combined in accordance with the requirements of the site, resulting in a plant design that aligns with the functional characteristics of the location.

Sustainable design method of landscape planting
Plants are a naturally sustainable resource, so it is an important sustainable design method to use plants that can adjust to the changes in climate, humidity, and temperature in urban spaces. Forest parks, wetland parks, rain gardens, roof gardens, green landscape walls, and natural habitats are common forms. From the perspective of urban planning, maintaining green spaces and ecosystems in regions, forest parks, and wetland parks is an important way to help the regional climate. At the same time, vast green planting spaces can accommodate thousands of plant species, protect rare tree species, and form a stable animal ecological community, which is of great significance to the ecological system.

In the era of limited urban land availability, urban green spaces have acquired an unprecedented degree of preciousness. A single recreational green space is insufficient to cater to the exigencies of contemporary urban landscapes. By incessantly exploring diverse forms of landscape planting, spatial configurations characterized by varying urban densities can enhance the sustainability of landscape greening. Among these, rain gardens emerge as the primary line of defense in the purification of urban road systems. When rainwater traverses road surfaces, it inevitably carries deleterious substances and residual particles. Discharging such water directly into adjacent rivers and oceans sans filtration exacerbates marine pollution and jeopardizes marine life. Therefore, an increasing number of countries advocate for the adoption of rain gardens as a means of enhancing municipal greenery. By optimizing the incline of municipal plantings and incorporating a combination of low-maintenance sedum planting[44], this choice of planting obviates the need for excessive roof and wall cladding thickness, rendering it highly adaptable to diverse structural configurations of roofs. Concurrently, an increasing number of projects prioritize the design and development of roof

密度的空间形态下都具有较强的可持续发展能力为目标。其中，雨水花园是城市道路系统的第一道净化防线。雨水冲刷道路表面后携带有害物质和残留颗粒，如果没有经过过滤直接汇入周边的河流和海洋，会增加海洋的污染和造成对海洋生物的伤害。越来越多的国家提倡进行雨水花园化的市政绿化改善，优化市政种植绿地的坡形，选择低维护及具有净化污水功能的植物组合。建筑屋顶是大量可更新使用的场地界面，对屋顶进行绿化改造，可以增加城市绿植的面积，同时起到调节区域微气候的作用。屋顶绿化和垂直绿墙通常选择耐旱低维护的景天属植物 [44]，这种植物对屋顶和墙面覆土厚度的需求较低，具有较强的可复制性，可以应用到不同结构和形态的屋顶上。同时，越来越多的项目注重屋顶界面的设计开发，把绿色种植

和休闲活动结合到屋顶上，形成极具视觉体验的空间场所。

自然风格的种植方式从 19 世纪的英国开始盛行。英国的温带海洋性气候形成了大片的森林和牧场，有着丰富的天然景观资源，从而形成了风景式园林的风格。但如绘画一样的"风景"为了保持其特定的构图形态和精美场景，往往需要大量的人力进行维护和修整。如此费时费力的景观造型方式显然与当下节能低维护的需求有着明显的冲突。尽管如此，人们对于自然的向往，以及对自然美的追求不会停止。荷兰景观设计师皮特·奥多夫 [45] 则运用自然野趣，把不需要过多维护的野生植物重新组合栽培，形成别具一格的景观（图 2-14）。世界各地的植物设计学已经开始探索低维护的植物营造方法，并大量运用在城市景观种植中，力

interfaces, integrating planting and recreational activities on rooftops to create visually captivating spaces.

Since the nineteenth century, naturalistic cultivation has enjoyed immense popularity in Britain. The mild maritime climate of Britain has given rise to vast expanses of forested and pastoral landscapes, which abound in natural forest landscape resources, thereby forming a garden landscape style. To preserve the specific compositional form of painted landscapes and their resplendent scenery, a significant amount of effort is required for their maintenance and restoration. This laborious and time-intensive approach to shaping the landscape evidently clashes with the contemporary necessity for energy-efficient and low-maintenance landscaping. Nonetheless, the yearning for nature and the pursuit of natural beauty remain ceaseless. In this regard, the Dutch landscape architect Piet Oudolf[45] has ingeniously employed wild plants, which necessitate minimal upkeep, to recreate a unique and natural landscape. Plant design methodologies across the globe have commenced exploring low-maintenance approaches to plant cultivation and are

being extensively employed in urban landscape planting. The objective is to achieve a sustainable urban landscape ecology that closely approximates the natural environment while conserving energy and water resources (Figure 2-14).

Saitama Sky Forest Plaza

图 2-11　日本埼玉新都心站"空中森林"广场

Habitat garden in four seasons

图 2-13　四季生境花园

Main tree with plant wall

图 2-12　主景树与植物墙

The Oudolf Garten at the Vitra Campus, Germany

图 2-14　皮特·奥多夫花园，德国莱茵河畔魏尔镇维特拉园区

求城市景观生态接近自然环境，并达到节能、
节水的可持续效应。

小结

　　自然环境拥有无限的潜力，我们可以持续
看到它的魅力和神奇。景观设计师在应对时代
的环境危机时，更需要展现对自然环境的学习
能力。通过科学和敏锐地观察自然环境，不断
在自然环境和人居环境两方面探索可持续设计
的可能，更好地为地球的环境生态做出贡献。

Summary

The natural environment presents boundless possibilities
for us to behold its charms and marvels. In an era marked
by environmental crises, landscape architects must
demonstrate their capacity to understand the natural
environment. Through scientific efforts and observations of
the natural environment, landscape architects continue to
explore the potential for coexistence and sustainable design
between the natural and human environments, thereby
making invaluable contributions to the environmental
ecology of the Earth.

课后作业
After-class exercises

讨论题
Discussion

气候环境的特征如何影响设计决策？

How do the characteristics of the climatic environment affect design decisions?

设计实践
Design Practice

针对全球气候变化，应解决建筑景观和硬质景观的哪些问题？根据讨论题内容，结合相关案例进行汇报论述。

What issues should be addressed in architectural landscapes and hard landscapes in the face of global climate change? Report and discuss with relevant case studies.

推荐读物
Reference reading

1. Barber, D. A. *Modern Architecture and Climate: Design before Air Conditioning*. Princeton University Press, 2023.
2. Cook, D. I., U.S. Forest Service, and Van Haverbeke D. F. *Trees and Shrubs for Noise Abatement*. University Press of the Pacific, 2004.
3. Dekay, M. and Brown, G. Z. *Sun, Wind, and Light: Architectural Design Strategies*. John Wiley & Sons, 2013.
4. Givoni, B. *Man, Climate and Architecture*. Van Nostrand Reinhold, 1981.
5. Lechner, N. *Heating, Cooling, Lighting: Sustainable Design Strategies Towards Net Zero Architecture*. John Wiley & Sons, 2021.
6. Olgyay, V. *Design with Climate: Bioclimatic Approach to Architectural Regionalism*. Princeton University Press, 2015.
7. Reed, C. *Retooling Metropolis: Working Landscapes, Emergent Urbanism*. Harvard University Graduate School of Design, 2017.
8. Roesler, S. *City, Climate, and Architecture: A Theory of Collective Practice*. De Gruyter, 2022.

第三章

城市与人工环境认知

Chapter 3
City and artificial environments

学习要点
Study Points
认识城市空间和城市规划的特点以及设计原则
Understand the characteristics and design principles of urban space and planning

学习目的
Teaching Objectives
认识城市和人居环境的设计关系
Understand the design relationship between cities and human settlements

在第二章我们认识了自然环境，在本章我们将开启对人居环境的认知。城市是将人、建筑、景观、生活、交通连接在一起的载体。我们将从城市的起源、城市的规划、城市的景观、城市的场所营造和社区建设等方面进行介绍，让大家对人居环境形成更充分的认知。

In this chapter, an analysis of urban networks of varying scales shall be undertaken. The city serves as a connection, linking people, buildings, landscapes, lives, and transportation. The origin of the city, its planning, the configuration of its landscapes, the establishment of its place, and the formation of its communities shall be expounded upon, thereby fostering a more comprehensive understanding of the living environment.

一、城市规划

本节通过城市的兴起、城市的扩张、城市的挑战三个方面，阐述城市规划对于城市可持续发展的重要性。

1. 城市的兴起

古罗马城是西方古典城市的集大成者。"伟大的艺术家创造了城市，而居民本身如果懂得该怎样生活于其中，就也是艺术家了。"[46] 罗马帝国曾是世界上城市系统发展最为发达的国家之一，集中表现在城市空间、功能组织、建筑美学的紧密协调上。这种魅力汇聚于城市广场、市场、庭院、街道、神庙等生活场所，这些场所不仅为当时的居民提供了身体的庇护，其所营造的精神世界更满足着居民对文化生活的追求。这些城市也被看作城市艺术，其建造者不只是具有城市规划背景的设计师，雕塑家、油画家、诗人、建筑师等也参与了城市设计。古罗马的街区、基础设施、建筑艺术为当代城市设计留下了经典的案例。其规划比例宜人的街道、广场和庭院空间，一直都是欧洲国家引以为豪的遗产，即使在今天，也值得我们借鉴和学习。

庞贝古城遗迹[47]给我们展示了十分完整的古罗马城市结构，包括交通网络、卫生基础设施、宗教建筑、休闲娱乐场所等完善的基础功能。庞贝古城的城市基础设施之先进，具体表现为成熟的交通、排水、供水系统。厚重石灰砖铺砌的街道可供行人及马车通行，而街道下层连接着分布在建筑周边的管道，浴场和生活废水都可以通过管道直接排入下水沟。高耸的连廊水道则是城市饮水、喷泉、生活用水的主要供水设施。完备的基础设施布局是古罗马城

3.1 Urban planning
3.1.1 The origin and development of cities

Ancient Rome, as the earliest archetype of a city, epitomized this notion. "Great artists create cities. If citizens know how to live in cities, they are artists themselves."[46] Ancient Rome once stood as the most advanced urban system worldwide, as evidenced by the close coordination of urban spaces, functional arrangement, and architectural aesthetics. This charm appeared in the squares, markets, courtyards, streets, fountains, sculptures, temples, and other habitation sites in the city. The spiritual realm created by these sites not only provided the residents with physical refuge but also instilled in them a yearning for religious culture and the pursuit of life. These cities were also regarded as urban art, and the creators thereof were not solely urban planning scholars but also sculptors, painters, poets, and architects who actively participated in urban design. The blocks, infrastructure systems, and architectural artistry of ancient Rome have left a typical example for the study of contemporary urban design. Even in the present day, centuries later, it remains a subject worthy of analysis and understanding. The harmonious and proportionate planning of streets, squares, and courtyards has consistently served as an intangible heritage that European countries take pride in.

The ruins of Pompeii[47] afford us the most comprehensive vestige of the urban fabric of ancient Romans. These remnants consist of urban transportation networks, public health infrastructure, religious buildings, recreational and leisure areas, and other fundamental facets of a city. The intricacy of Pompeii's urban infrastructure is reflected in its transportation systems, urban drainage, and water supply. Streets paved with thick lime bricks facilitated the passage of pedestrians and wagons alike. Subterranean pipes interconnected with the surrounding buildings, enabling the direct discharge of bathroom and domestic sewage into the lower drainage channels. The towering aqueduct served as the primary source of potable water, fountains, and domestic water supply in this city. The layout of these infrastructural elements constituted the most sagacious aspect of ancient Roman urban planning, thereby establishing the bedrock for the city's prosperity and stability.

市规划中最具智慧的设计，为城市的安定和繁荣奠定了基础。

在中国，古长安城（今中国西安）也是同时期设计最为先进的城市之一。[48]整体的规划布局以中轴对称为主，以网格式的布局分割城市，每一个网格为一个坊里，类似今天我们说的"街区"。[49]在城市结构的划分中，以皇宫为中心，分为东西两市区，同时规划了园林、寺庙建筑、集市等公共空间。繁华的长安城曾是东方世界的中心、"丝绸之路"[50]的起点，也是古罗马、波斯、叙利亚等远邦进行商业和文化交流的集中地。

从中外两个古城的城市设计可以看出，城市生活、经济、文化的发展与城市的基础设施配置、公共空间品质和交通道路规划息息相关。最宜人的城市都是对其所处时代、地域和文化的综合阐释和回应。我们在进行城市公共空间的设计时，也要从古老的经典城市规划中寻找价值点，理解广义和狭义的城市美，从而选择更优的设计策略。

2. 城市的扩张

到了 19 世纪初，随着人口不断增长，工业革命后的环境污染危机初现，大批量乡村劳动力涌入城市，城市在扩张的同时也迎来了许多问题。[51]

第一，城市的扩张造成环境破坏。大量的住宅项目开发造成景观资源的破坏，其中包括地质侵蚀、土壤污染、植被过度砍伐、野生动物消亡等问题。城市的扩张需要大量建造楼房、高速公路、生活基础设施等，这会对自然环境的生态系统产生侵害。例如，对郊区和乡村的农田、野生栖息地等绿化空间的征用开发，过

The ancient metropolis of Chang'an (present-day Xi'an, China)[48] stood as a paragon of sophistication in Chinese urbanism. Its architectural layout boasted a symmetrical configuration, centered around a prominent axis, with urban blocks divided into grid patterns akin to contemporary "blocks."[49] The urban structure was demarcated by the palace's central domain, which was further subdivided into eastern and western urban sectors. Additionally, the city designed garden areas, temple buildings, bustling markets, and various other public spaces intended for recreational pursuits. Notably, Chang'an served as the nucleus of the East and the originating point of the Silk Roads [50], facilitating artistic and cultural exchanges between distant lands such as ancient Rome, Persia, and Syria.

The urban design of these two ancient cities highlights the inextricable link between urban infrastructure allocation, the quality of public spaces, and the reasonable planning of roads in fostering the development of urban life, economy, and culture. "To a great extent, the most livable city should be the best expression and response to the characteristics of the time, regions, and culture: multi-functional, accessible, reasonable, and complete."

In contemporary urban public space design, it is necessary to discern the inherent value in the annals of classical urban planning and to grasp the essence of urban aesthetics, thereby enabling the selection of optimal design strategies.

3.1.2 Urban expansion

By the mid-nineteenth century, the burgeoning population, coupled with limited resources, the environmental predicament after the Industrial Revolution, and the influx of rural labor into urban centers, led to a host of challenges that accompanied urban expansion.[51]

First and foremost, it is necessary to acknowledge that land development has resulted in significant environmental degradation. The ramifications of extensive residential development involve geological erosion, soil contamination, rampant deforestation, and the depletion of wildlife populations. The expansion of urban areas has precipitated the construction of numerous

程中造成了生态环境的污染。大面积的湿地受到破坏，将会影响水资源和生物赖以生存的栖息环境。就像美国城市奥兰多的扩张，因为缺少对自然环境的保护和扩张的限制，使湿地面积不断减少，造成本地生物链的严重破坏和气候温度的上升。

第二，城市的扩张引起经济动荡。楼房的大面积开发，也会形成房地产经济的不均衡布局。这也是我们所看到的，越是商业活动频繁的区域（市中心），房价也相对较高的现象。而大量居民因为不能承受的房价，搬到通勤路途较远的副中心甚至郊区生活。人口往城市郊区迁移，这种现象会导致市中心缺乏活力，同时也需要更多的汽车和高速公路来满足市民日常生活的出行，从而产生更多的污染排放。

第三，城市的扩张带来安全隐患。美国著名的城市洛杉矶以好莱坞造梦工厂闻名，但另一方面，这个城市也被叫作"日落犯罪之城"。[52] 在早期开发的过程中，洛杉矶出现了哄抬市中心房价的现象，导致大量居民搬离城市中心。市中心的建筑在工作日办公时间以外的时间段多处于空置的状态，在晚上又会聚集很多非法移民者和穷困住户，导致犯罪率不断攀升。

3. 城市的挑战

面对环境的资源限制与存量时代的土地开发问题，城市规划已经从大拆大建，走向修缮与更新。越来越多的城市在思考如何加强对可再生资源的利用。首先，通过技术发展减少不可再生资源的使用。新加坡一直致力于再生水资源 [53] 的发展，利用海水转换淡水技术，满足新加坡的城市日常供水需求。不仅如此，政府还通过奖励补贴，鼓励开发商建造绿色建筑与

buildings, highways, and residential infrastructure, thereby encroaching upon the natural environment. Notably, the conversion of agricultural land, pristine habitats, and verdant spaces in suburban and rural regions for developmental purposes has resulted in ecological pollution. The obliteration of vast expanses of wetlands has adversely impacted water resources and the ecosystems reliant upon them. A case in point is the urban sprawl witnessed in Orlando, USA, where the absence of environmental protections and constraints on expansion has precipitated a steady decline in wetland areas, extensive disruption of the ecological chain, and an exacerbation of climatic temperatures.

Secondly, the expansion of urban areas has led to economic upheaval. In addition to the development of buildings, urban expansion has resulted in an imbalanced real estate economy. The exorbitant property prices have compelled a substantial number of residents to relocate to areas situated further away from their workplaces. The surge in property prices has impeded the migration of individuals from urban centers to the outskirts of the city. This

phenomenon has resulted in a dearth of vibrancy in the city center and has necessitated the expansion of automobiles and expressways to cater to the daily commuting needs of citizens, thereby exacerbating pollution levels.

Thirdly, urban sprawl has precipitated concerns pertaining to urban security. Los Angeles, USA, colloquially known as "the city of angels," has also earned the moniker of "the sunset city of crime." [52] The early stages of Los Angeles' development resulted in inflationary pressures in the downtown area, prompting a significant exodus of residents to more distant locations. Therefore, the city center remains desolate outside of office hours on weekdays. During the nighttime, the tranquil streets turn into a haven for criminal activities.

3.1.3 Urban challenges

In the face of environmental resource constraints and land development issues in the era of limited urban land availability, a shift has occurred from large-scale demolition and construction towards urban restoration and renovation. Increasingly, cities are contemplating strategies for

垂直绿化的项目，从而在建筑扩张过程中减少碳排放。垂直绿化让国土面积狭小的新加坡快速得到人均 8m² 的绿化率 [54]（图 3-01）。在城市发展的过程中通过设计规划调整城市绿地分布格局，也能让城市功能结构更为合理。

巴塞罗那在 20 世纪 80 年代的旧城改造中，对有价值的历史建筑、老城区街道、商业街区和城市广场进行联合改造，通过调整城市业态让旧城区建筑的文化价值得到进一步提升，从而促进城市的复兴。城市的更新并不是单一的"新形象"展现，更是城市文化的传承。如图蒙特卡达街道上的斯托达·诺瓦新型市镇别墅、费兰和公主商业街道、新的纪念性广场如圣若梅广场（图 3-02）、莱埃塔那大街（图 3-03）、兰布拉大道（图 3-04）、巴塞罗那道路系统（图 3-05）等。美国的纽约高线公园，同样使用城市复兴的策略，把旧有的铁路空间置入新的城市文化中。高线公园与其他城市公园相比，并不是传统的绿化公园，而是激活城市废弃场所的文化公园。在高线公园改造前，切尔西地区是滨水区域的工厂用地，用于旧时的码头货运。随着城市中心化的发展，该滨水工厂用地一直处于无人往来地区，周边商业也并不活跃。但高线公园利用不一样的视线角度，通过"特殊城市景观"的观察方式进行公共空间的休闲设计，对现代都市的"绿地"形式进行了新的阐述，从而带动了整个片区的商业经济发展，甚至成为纽约乃至世界的经典地标。

每个城市的挑战各不相同，也各有相似。每个国家都在寻找与本土文化契合的设计方式，以解决城市发展和自然环境的冲突，为更好的未来环境而努力。

harnessing renewable resources in urban development. Firstly, advancements in technology have facilitated a reduction in the utilization of non-renewable resources.[53] Singapore, for instance, has dedicated itself to water resource development and is actively employing recycling technology to convert seawater into freshwater, thereby ensuring an adequate urban water supply. In addition, the government incentivizes developers to construct eco-friendly buildings and implement vertical greening initiatives, effectively reducing the carbon footprint associated with urban expansion. Vertical greening enables the rapid greening of limited public spaces at a green ratio of 8 square meters per capita[54] (Fig 3-01). Additionally, the design and planning of urban development processes can optimize the functional structure of cities by adjusting the distribution of green spaces. In the 1980s, the Spanish city of Barcelona underwent a transformation by integrating historical buildings, streets, business districts, and city squares. This restructuring of the urban fabric not only enhanced the architectural and cultural value of the old city but also contributed to urban regeneration. Urban renewal serves not only as a reflection of a "new image" but also as a conduit for the transmission of urban culture (Fig 3-02–3-05). In the case of New York City's High Line Park, a similar strategy of urban regeneration was employed to repurpose an old railway into a novel urban space. Diverging from conventional city parks, High Line Park functions as a cultural park that revitalizes abandoned urban areas. Prior to the park's renovation, the Chelsea District housed a former waterfront factory that served as a cargo transportation hub at the old pier. As the city center developed, this coastal factory area became inaccessible, resulting in a lack of commercial activity in the vicinity. However, High Line Park, as a public space, was designed to provide recreational opportunities through a "special urban landscape." This innovative interpretation of "green space" in modern cities has fostered the economic growth of the entire region and has become an iconic landmark in New York City and beyond.

While cities face diverse challenges, they also share commonalities. Each country seeks culturally appropriate design methodologies to address the coexistence of urban development and the natural environment, striving to forge a more sustainable environmental future.

Parkroyal Collection Pickering, Singapore

图 3-01　新加坡花园酒店

Plaça Sant Jaume, Barcelona

图 3-02　巴塞罗那圣若梅广场

Via Laietana, Barcelona

图 3-03　巴塞罗那莱埃塔那大街

图 3-04 **La Rambla, Barcelona**
巴塞罗那兰布拉大道

图 3-05 **Road system update in Barcelona**
巴塞罗那道路系统更新

二、 城市的特征

本小节根据城市的属性分类，从城市结构形式、空间成分、基础设施网络构成等方面介绍城市的内涵和特征。

1. 城市的结构

理解城市的结构有助于我们了解城市是如何运作的。我们需要明白城市任何一个地方的设计与改造，都不是独立的单体存在，而是与整体城市系统和环境气候紧密相连。大部分城市呈现环状布局、散装布局、棋盘式布局等，但是都可以从空间上划分为中心城区、内环城区、外环城区以及郊区林地。中心城区，也被称为中央商务区（CBD），是每个城市政府、商业和金融机构最为密集的地区。市中心以极具密度的商业文化活动展示城市最有活力的一面。历史建筑、艺术表演中心、博物馆、图书馆、画廊、商业街道、咖啡厅、餐馆等是其周边建筑的主要业态。市民和游客都会聚集在市中心的商业广场、街道市集、街角绿地、户外餐饮区等进行活动，丰富的商铺和娱乐空间为中心城区带来了街区的活力。市中心大多属于城市租金最高且停车空间较为紧张的区域，特别是历史悠久的市中心，还保留着古老的城市街道结构，不便于车辆的通行。市中心的交通网络多以步行道为主，并依赖公共汽车为主要的接驳车辆。例如，多伦多、巴黎、罗马、佛罗伦萨的市中心就通过控制道路车辆数量和停车位置来限制中心的车流量，从而保证道路不受车辆的影响，形成步行友好的街道空间，让居民在中心区感受商业娱乐活动带来的街区生活活力。部分大型城市中心无法完全截断汽车的通行，会利用地下通道让汽车快速通行，缓

3.2 Characteristics of cities

In this section, we explore the essence and characteristics of cities by categorizing them based on attributes such as urban structural forms, spatial components, and the composition of infrastructure networks.

3.2.1 Urban structure

Understanding the urban structure is necessary for understanding its functionality. It is crucial to acknowledge that the design and alteration of any urban segment do not exist in isolation but rather exhibit a profound interconnection with the entire urban system, as well as the surrounding environment and climate. While most cities possess a circular and loosely connected layout, they can be categorized into the central city, inner city, outer city, and suburban areas. The central business district (CBD) represents the most bustling region in each city, accommodating governmental, commercial, and financial institutions. This district comprises historical buildings, performing arts centers, museums, libraries, galleries, commercial thoroughfares, cafes, and restaurants, which serve as the primary commercial enterprises in the vicinity. The downtown blocks are invigorated by a plethora of shops and entertainment establishments, fostering a vibrant atmosphere. The downtown commercial square, street market, corner green spaces, and al fresco dining areas serve as gathering spots for both residents and tourists, facilitating leisurely activities.

Most city centers are situated in urban areas characterized by exorbitant rental prices and limited parking spaces. Particularly in city centers, the traditional street structure of the old city is often preserved, rendering vehicular access arduous. Therefore, the transportation network in the city center predominantly caters to pedestrians, with public transportation serving as the primary mode of transit. For instance, urban centers such as Toronto, Paris, Rome, and Florence regulate traffic flow and parking availability to restrict vehicular volume in their respective cores. This approach ensures that thoroughfares remain unencumbered by vehicles, thereby creating pedestrian-friendly street spaces that enable residents and visitors to immerse themselves in the vibrant ambiance of the

解市中心的车流压力。例如，纽约中央公园设置地下车道让车辆快速跨区，上海外滩中心设有地下车道通往滨江和黄浦江对岸。

内环城区是由大量的居住和生活设施形成的区域。内环城区较早被开发为成熟的居住社区，城市配套服务如医院、学校、文化康乐设施等都较为完善。同时，内环城区为在市中心工作的居民提供与高低中收入相匹配的不同住房生活区，为城市的更新发展提供人才支持。但内环城区也存在老旧住宅和空置厂房较多的问题，因开发或修缮成本较高，开发商宁愿在郊区兴建新的住宅区也不愿意投入更多的成本改造旧建筑。内环城区在原有的城市基础设施基础上应加大发挥城市更新的力量，合理规划不同收入居住群体的城市住房，为中心城区和外环发展提供人力保障。

外环城区是市中心和内环的扩充区域，通常以较大尺度的卫星城规划为主。卫星城与中心城区以高速公路和绿色环道连接而成，目的是从中心城区能够较快速地到达卫星城区。外环城区多以加工制造、医疗保险、商业办公、娱乐等设施为主，同时配有大量的员工居住社区和新移民小区。外环城区通常是集有效的基础服务设施、绿地休闲景观、生活娱乐配套于一体的综合社区。郊区及林地，可以理解为最接近自然生态的区域。该片区居住人口较少，以从事农业、畜牧业、厂房原材料加工的人群为主。郊区因为缺少集中的生活基础设施更新，以及年轻劳动力的损失，其商业和居住吸引力在不断降低，从而形成破败简陋的建筑形态。但从自然保护的角度看，郊区的林地和未遭破坏的郊野生态是十分珍贵的自然资源。对于郊

neighborhood. In certain instances, large urban centers cannot entirely eliminate vehicular traffic and thus incorporate subterranean passages to facilitate the movement of cars through the city center. Notably, New York's Central Park features an underground roadway that expedites vehicular transit, while Shanghai's Bund features a subterranean thoroughfare that grants access to the riverside area and the opposite side of the Huangpu River. The city center epitomizes the most dynamic facet of urban life, teeming with commercial and cultural activities.

The inner city comprises a multitude of residential areas and living facilities, constituting a mature residential community. In this urban domain, an array of services, hospitals, schools, and cultural and entertainment amenities have been established, readily accessible to its inhabitants. Simultaneously, the inner city offers diverse housing and living options for individuals employed in the city center, thereby bolstering the city's rejuvenation efforts. Nevertheless, the inner city is burdened with numerous vacant antiquated dwellings and factories. Due

to the exorbitant costs associated with their development and maintenance, developers exhibit a preference for constructing residential areas in the suburbs, rather than investing additional funds into renovating existing structures. The vacancy rate in the city center is inversely correlated with the demand and supply of housing, precipitating a substantial moving out by families to the outskirts or suburbs, as a means to alleviate the pressures of urban life. Leveraging the existing urban infrastructure, it is necessary for the inner urban areas to harness the potential of urban renewal, aggregating residential clusters and formulating reasonable urban housing plans, thereby fostering the advancement of requisite services and facilities in the central city and its periphery.

The outer city represents an expansion of the inner city and the city center, typically conceived as a larger satellite town. These satellite towns are interconnected with the central city through expressways and verdant ring roads, facilitating swift transit from the central city to the satellite towns. Within the outer ring, satellite towns predominantly include manufacturing, medical,

区的开发和更新，不能一味地复制开发市中心和内环区域的规划策略，更应该考虑农业、村庄、森林、原始保护区相结合的旅游产业业态，从而增加本地居民的工作机会，也为城市片区生态发展提供保护。

2. 城市的空间成分

交通网络是城市构成的主要骨骼，景观公共空间和建筑就如同身体器官的各个部分通过交通网络被连接在一起。交通网络按使用特点可分为城市道路、公路、厂矿道路、林区道路、乡村道路等。城市道路又可细分为人行道、车行道、城市快速道路、城市主干道、城市次干道、城市支路，以及胡同和里弄等。设计道路环境需要根据城市周边的生态系统、植物区域、水文格局、道路宽度和预计的交通流量进行综合估算。道路的分布影响着区域的经济发展，

交通网络布局合理及高效的区域能吸引更多的公司进行办公置业，从而促进就业率和房地产的区域升级。但同时也要注意，不断扩张交通道路会增加道路边缘杂草和外来有害物种的入侵，并且雨水冲刷带来的淤泥、铅、盐等对水系统的污染较大。高速公路的修建也会造成草地鸟类的减少，以及河流鱼类栖息地的减少。

在城市空间中，建筑是主要的人居栖息场所，主要分为四大类。第一类以居住商用为主，具体为居住用地、行政办公用地、教育科研用地、社会服务用地、商业用地、商务用地。第二类为基础设施用地，具体包括公共设施营业网点用地、公共交通场站用地、供电用地、服务设施用地、文化设施用地、医疗卫生用地、文物古迹用地、其他服务设施用地。

第三类为交通服务空间，具体包括社会停

and business offices, as well as entertainment venues, alongside a large number of residential communities and new residential areas. The outer ring epitomizes a novel comprehensive urban planning community, effectively integrating fundamental services, verdant spaces, and living and recreational amenities. The suburbs, in turn, can be perceived as areas in the closest proximity to natural ecology. The population residing in these environs is relatively modest, predominantly comprising individuals engaged in agriculture, animal husbandry, and raw material processing for factories. The absence of centralized urban living infrastructure, coupled with the depletion of a youthful labor force in the suburbs, has resulted in a decline in the appeal of businesses and residences, resulting in dilapidated structures and substandard quality. However, from the standpoint of nature conservation, the suburban green spaces and rural ecology represent invaluable natural resources. The development and revitalization of the suburbs should not adopt the same planning strategies employed in the city center and inner ring. It is critical to contemplate the integration of agriculture, villages, forests, and existing protected areas, with a view to fostering tourism, enhancing residents' employment prospects, and protecting the ecological progress of urban areas.

3.2.2 Spatial composition of the city

The transportation network constitutes the primary framework of the city, interconnecting the public spaces and buildings of the urban landscape. It comprises various types of thoroughfares, namely urban roads, highways, forestry roads, and rural roads. Urban roads can be further categorized into sidewalks, carriageways, urban expressways, urban main roads, urban secondary roads, urban branches, and hutongs. The design of the road environment is predicated upon a comprehensive assessment of the surrounding urban ecosystem, vegetation coverage, hydrological model, road width, and projected traffic volume. The spatial distribution of roads exerts a profound influence on the economic development of a region. A reasonable and efficient layout of the transportation network attracts a greater number of enterprises to establish their presence in the area, thereby enhancing employment opportunities and enhancing the quality of real estate. Nevertheless, it is necessary to

车用地、消防用地。

第四类为专属用地类型，具体包括体育用地、体育训练用地、特殊医疗用地、战略性部署用地、外事用地等，此类型用地可根据本土规划需求而进行专项用地使用。

当代的建筑设计需要对城市、景观空间进行统一规划与设计，建筑设计不是孤立的存在，而是与环境相互联系，形成紧密关联的空间形态。如伦敦奥林匹克体育场，建筑的场馆设计具有可拆卸、可持续利用的功能，为城市空间发展提供节能减能的效果（图3-06）。

景观公共空间包括位于交通网络和建筑物之间的各种城市公共区域，可划分为庭院空间、广场空间、高层景观、屋顶景观、步行绿道、公园绿地、水域等。庭院空间划分大型建筑空间中的主要休闲活动场所，丰富的绿植点缀和

艺术化的表达让建筑的流线富有变化，同时增加室外的观赏点。如日本庭院，通过四季绿植的变化，形成一道道亮丽的风景线。美国耶鲁大学图书馆内庭院，通过雕塑化的几何形式营造，为室内阅读环境带来步移景异的效果（图3-07）。广场空间是城市中最为吸引人的场所。节假日的休闲活动等不同类型的文化活动都可以在这里举办。它展示了城市独特的魅力，也体现了城市的历史人文气息。如意大利的特雷维喷泉[55]，依据神话故事塑造了阿波罗的精美雕像，尽管时代变迁，这里一直是意大利以及世界游客喜爱的场所。德国索尼大厅广场，线性的几何风格奠定了广场简约的设计基调，在周边用餐、购物、办公的人都可以在广场里找到适合自己休息的地方。

屋顶花园和高层景观是近年来城市景观的

acknowledge that the incessant expansion of roads can lead to the encroachment of weeds and pernicious species along the road periphery. Moreover, sediment, lead, and salt generated by rain erosion can contaminate water systems. The construction of expressways can also reduce the population of grassland avifauna and encroach upon the habitats of riverine fish.

In the urban space, architecture can be classified into four principal categories. The first category primarily comprises residential and commercial domains, including residential land, administrative office land, educational and scientific research land, social service land, business land, and enterprise land. The second category pertains to infrastructure-related land, including land for public facilities and commercial establishments, land for public transportation stations, land for power supply, land for service facilities, land for cultural amenities, land for medical and healthcare facilities, land for cultural heritage sites, and land for other service facilities. The third category comprises spaces associated with transportation, including social parking lots and fire-fighting land. The

fourth category includes proprietary land use, including sports land, sports training land, specialized medical land, strategic deployment land, and foreign affairs land. Such land can be utilized for specific purposes in accordance with local planning requirements. Contemporary architectural design has become integrated with urban and landscape planning and design. Architectural design is no longer an isolated entity but rather an integration with the environment, giving rise to an interconnected spatial configuration. For instance, in the Olympic Park of London, the design of pavilions and buildings incorporates features that render them removable and sustainable, thereby facilitating energy conservation and emission reduction (Fig 3-06).

Landscape public spaces encompass a variety of urban public areas situated between the transportation network and the buildings. These spaces include courtyard spaces, plaza spaces, high-rise landscapes, rooftop landscapes, pedestrian greenways, parklands, and water areas. The courtyard space, situated within a large building, serves as the primary recreational area. Its abundance of greenery

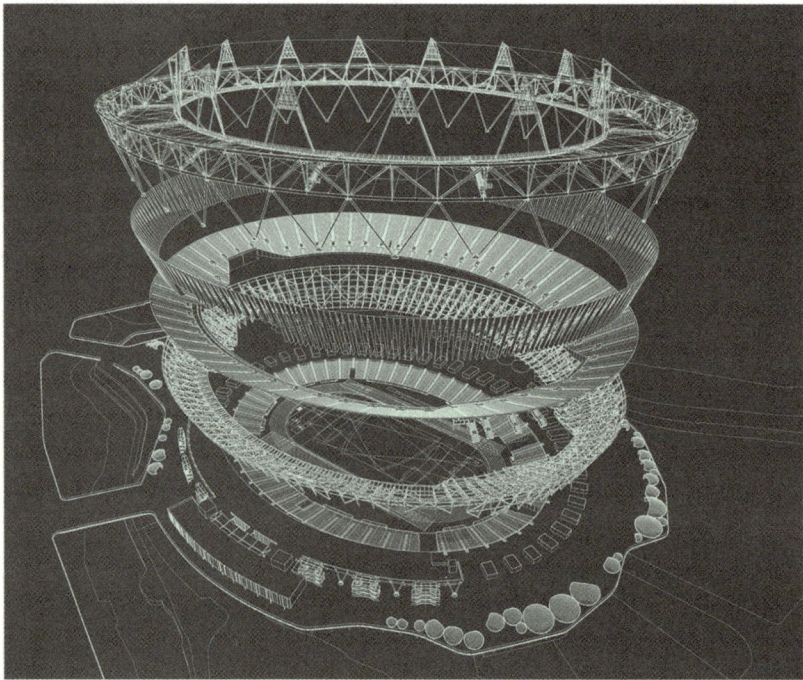

London Olympic Stadium taken apart

图 3-06　伦敦奥林匹克体育场拆解图

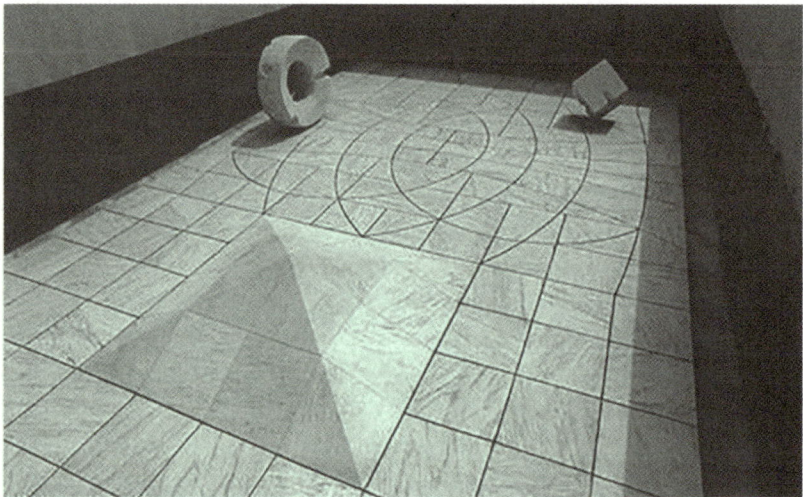

Yale University Library

图 3-07　耶鲁大学图书馆下沉庭院

一大发展趋势。随着城市用地的紧张，越来越多的建筑在寻求绿色低碳的设计方法。而屋顶花园的建造形式和节能技术已经十分成熟，可以巧妙地连接建筑空间，形成高层景观的独特表现形式。在高空中呈现热带雨林般的景观步道，或者在屋顶花园设置无边际游泳池，都可以创造高空独特的体验视角。步行道、公园绿地、水域滨水空间是城市日常生活重要的景观活动区域。步行道是交通道路旁具有统一性和指向性的绿植廊道，同时提供遮阴挡风功能。公园绿地则是城市的"绿肺"，对调节城市区域微气候，保护生物物种有着重要的战略性意义。水域滨水空间是城市景观元素中的重要保护资源。拥有江、河、湖泊、溪流的城市，都有着丰富的植物和水生物资源。水域也是重要的淡水资源，对城市的用水起到重要的作用。

这要求城市保护水资源流域免受污染。滨水空间的规划更是城市设计中不可忽视的部分。

3. 城市中的蓝绿基础设施网络

植物与水都有助于优化城市生活的环境，这也形成了景观设计中的蓝绿系统。[56] 蓝色代表水系统（蓝色基础设施）设计，绿色代表植物（绿色基础设施）设计，两者结合形成共同发挥生态服务功能的空间网络。蓝绿基础设施强调共同营造的自然空间，以应对环境危机的挑战。例如，从水治理的角度来说，城市雨洪问题并不能因为自然河道防洪设施的修整而完全解决。城市人口增长，土地的开发缩减地表收集雨水的植被，导致城市的水流加速而形成大量的积水问题，这会进一步导致城市植物生态以及区域气候的变化。这种恶性循环必须通过蓝绿基础设施协调发展，增强城市对突发灾

and artistic expressions not only alter the building's flow but also enhance outdoor viewing points. Similar to Japanese gardens, the greenery in these spaces changes with the seasons, resulting in a picturesque landscape. A notable example is the library of Yale University in the United States, which employs sculptural geometric structures to introduce the effect of changing scenery into the indoor reading environment (Fig 3-07). The square, serving as a focal point in a city, holds immense attraction. It serves as a venue for public holidays, leisure activities, and various cultural events. The square showcases the city's unique charm while embodying its historical and humanistic atmosphere. For instance, the Fontana di Trevi[55] in Italy, features a captivating Apollo statue inspired by myth. Over time, it has become a beloved destination for tourists from Italy and around the world. Germany's Sony Concert Hall Square, characterized by its linear geometric style, laid the foundation for minimalist square design. Individuals working in the surrounding restaurants, shops, and offices find solace in this square, using it as a place to rest.

In recent years, rooftop gardens and high-rise landscapes have emerged as significant transformations in urban landscapes. As urban land becomes increasingly scarce, more and more buildings are seeking green and low-carbon alternatives. Rooftop gardens, already well-established in terms of construction and energy-saving technology, seamlessly connect building spaces and offer a unique expression of high-rise landscapes. For instance, the Singapore Green Corridor presents a tropical rainforest landscape walkway elevated above the ground, while some rooftop gardens feature infinite swimming pools, providing a unique aerial perspective. Pedestrian pathways, green spaces, and waterfronts play crucial roles in daily urban life. Sidewalks serve as unified and directional green corridors along roads, providing pedestrians with shade and protection from the wind. Green spaces, acting as the city's green lungs, hold strategic importance in regulating the microclimate of the region and protecting urban species. Waterfront spaces serve as vital resources for urban landscapes. Cities blessed with rivers, lakes, and streams feature abundant plant and aquatic life. Additionally, cities must protect these water bodies from pollution.

害的应变能力来改善。这也是我们经常提到的增强城市韧性的问题。在城市规划中，通过蓝绿基础设施的设计，模拟自然环境内外循环的生态功能，从而在基础设施层面上减少自然灾害对城市的影响。这些基础设施包括交通、水、能源、材料的可循环系统等。[57] 如透水铺装的应用，可减少排放到海洋的地表径流污染物；雨水花园和湿地生态环境的建立，可保护区域生态物种的平衡；建造屋顶花园和城市公园、减少停车空间等方法，都可以提高绿地比例，改善区域的冷却功能，降低热岛效应的影响。城市用水的供给、排放，以及湖泊海洋的水系统循环都是提高城市韧性、增强可持续发展的重要部分。如哥本哈根的蓝绿设施改造、意大利的垂直森林、荷兰的冰雪公园设计等，设计师都在城市空间中发挥了不同的空间创意，协调自然与非自然元素的内外联系，致力于创建可持续的韧性城市环境。

三、城市公园和绿地

本节内容通过介绍公园的起源和发展、公园的价值、公园的类型，分析城市公园和绿地的特征，理解公园和绿地对于城市和谐发展的重要性。

1. 公园的起源和发展

在古代，罗马人建造了柱廊运动场，并在旁边建造公共花园和郁郁葱葱的空间。古罗马人爱好竞技娱乐，这种回廊花园兼具运动的功能，拱形回廊则具有遮阴避暑的功能，也成为讲学及休闲的场所。据史料记载，古罗马大概建造了 30 座这样的花园向公众开放使用。在中世纪的欧洲，人口的密度不断增加，建筑多

Simultaneously, river basins serve as essential freshwater resources, playing a significant role in urban irrigation and water supply. Therefore, water planning becomes an integral part of urban design, inseparable from its overall framework.

3.2.3 Blue-green infrastructure network in cities
Plants and water play a crucial role in enhancing the urban living environment, thereby establishing the blue-green system in landscape design.[56] The color blue symbolizes the design of the water system (blue infrastructure), while green represents the presence of plants (green infrastructure). Together, these elements form an intricate ecological network. Diverging from conventional functional classifications of water systems and vegetation design, the blue-green infrastructure accentuates the creation of both natural and artificial corridor spaces, aimed at addressing the prevailing environmental crisis. From a water management perspective, the predicament of urban rainstorms remains unresolved due to the restoration of natural flood controls along rivers. The burgeoning urban population has led to encroachments on vegetation responsible for surface rainwater collection, therefore accelerating urban water runoff and exacerbating waterlogging issues. This, in turn, triggers changes in urban vegetation and regional climate, thereby perpetuating a detrimental cycle. To break free from this vicious circle, a coordinated development of a series of blue and green infrastructures is necessary, as it enhances the city's resilience against unforeseen disasters. This concept of bolstering urban resilience is frequently discussed.

The impact of natural disasters on cities has compelled urban planners to devise blue-green infrastructures, including transportation, water, energy, and material recycling systems[57], among other collaborative projects that are jointly planned and designed. For instance, the utilization of permeable paving bricks effectively curtails the discharge of pollutants from surface runoff into the sea. The establishment of rain gardens and wetland ecosystems protects the ecological equilibrium of regional flora and fauna. Roof gardens, city parks, and the reduction of parking spaces all contribute to enhancing the green space quotient, thereby enhancing the area's cooling functionality and mitigating the heat-island effect. Urban water supply,

以城堡、教堂和庄园为主。而城市的规模较小，所形成的娱乐及公共的场地也较少。教堂前的台阶广场和街道，是主要的公共空间，也是传道士宣讲和戏剧表演的场地。到了文艺复兴时期，庄园和私人庭院得到了大规模的发展。园艺学、雕塑工艺以及科学的造园艺术都在这一时期达到新的顶峰。而这些庄园、私人庭院、皇室宫殿一般不对外开放。直到工业革命以后，欧洲大批的私人庭院和宫殿才开放成为城市公园供市民使用。大部分皇室开放的宫殿，如圣詹姆士公园、摄政公园等，都是以运河、大片田园森林为特征的浪漫主义风格园林，具有较高的艺术价值，以模仿神话故事中的场景为主，形成与自然环境融合的美丽圣景。强调画面的园林虽然具有较大的观赏价值，但对公众的参与性缺乏考虑，所以即使开放了园林参观，也

没有在真正意义上转化为公众的公园。世界上第一个城市公园是英国的伯肯海德公园。建筑师约瑟夫·帕克斯顿拟定了利物浦区附近的公园和投资房产的草案。在该草案中，规划方案必须有 70 英亩（约 4.6 公顷）的公共绿地供居民进行休闲娱乐。居住楼房环绕式布置在公园四周，给予住宅无遮挡的绿化视野。在公园的规划中，丰富的地形形成模拟自然山水场景的森林和湖泊。中心一级车行步道可直通商业中心，二级步行环道可不受车辆的干扰，形成环状的休闲娱乐步道。伯肯海德公园的成功体现了自然环境对房产价值的提升，也开始让规划者意识到城市公园对于调节城市环境和市民活动的重要性，从而开启了城市公园兴建的繁荣时期，之后世界各地开始推广不同的城市公园。其中最为经典，影响最为深远的就是我们

drainage, and functional water systems are pivotal measures in improving urban resilience and fostering sustainability. Notable examples include the renovation of blue-green facilities in Copenhagen, the vertical forest in Italy, and the design of an ice and snow park in the Netherlands. In these instances, designers have skillfully explored diverse spatial possibilities in urban areas, aiming to enhance the relationship between natural and built elements, with the objective of creating a sustainable and adaptable urban environment.

3.3 City parks and green spaces
3.3.1 The origin of the park
During ancient times, the Romans utilized colonnaded sports grounds, alongside documented accounts of public gardens and verdant spaces. The Romans, known for their athletic prowess, repurposed these cloister gardens as sports arenas. The arcaded cloisters not only offered respite from the scorching sun and heat but also served as venues for lectures and relaxation. Historical records indicate that approximately thirty such gardens were accessible to the general public. As Europe transitioned

into the Middle Ages, population density surged, resulting in the prevalence of castles, churches, and manor houses in medieval architecture. Conversely, urban areas were relatively smaller in size, limiting recreational and communal spaces. The sole existing venues were the stepped squares and streets adjacent to churches, which functioned as primary marketplaces and sites for religious sermons and theatrical performances. The Renaissance era witnessed the widespread development of manors and private courtyards, where horticulture, carving techniques, and the art and science of gardening reached their zenith. These manors, private courtyards, and royal palaces were typically inaccessible to the public. It was not until the nineteenth-century Industrial Revolution that numerous private courtyards and palaces across Europe were opened to the public as urban parks. Prominent examples include St James' Park and Regent's Park, which embody romantic gardens featuring canals and expansive pastoral forests. These gardens possess remarkable artistic value, drawing inspiration from fairy tale settings and seamlessly blending with the natural environment to create enchanting and sacred vistas. However, the emphasis on patterned

熟悉的纽约中央公园，它把城市公园的理念植入城市规划设计，为纽约城市百年的发展不断提供生态保护和城市文化活力。

2. 公园的价值

我们以浓缩的公园史展示了城市公园的起源和发展，而城市公园对于城市未来绿色资源的发展有着重要的意义。公园的价值主要可分为环境价值、经济价值、人文价值三部分。环境价值，这部分不难理解。在之前的章节中，我们已经了解了城市的蓝绿系统对于调节城市的气候和生态环境有着重要作用。城市公园看似简单的绿化种植，但是其所形成的天然生态圈却时刻保护着我们的生存环境。例如，公园大面积的绿化和种植本地树种的林荫道，可在雨水期吸收大量地表废水，从而减轻城市的排洪压力。种植的植物所形成的天然廊道，可为城市减少区域的热岛效应。土壤层和植物树丛也为城市中的昆虫、鸟类等提供自然栖息地，形成稳定的生态活动圈层。经济价值，从规划的角度可验证，居住工作区域有着较好的城市公园环境，周边的房地产价值会有较快提升。人文价值体现在城市公园不仅提供自然健康的绿地环境，还提供大量的休闲活动空间，供人们进行运动、日常休息、大型聚会活动等。当城市的活力被公园绿地所激活，周边的娱乐场所、餐厅等生活服务类的设施也因为大量的人流和消费而变得繁荣。再次以纽约的高线公园为例，本来废弃而人烟稀少的切尔西社区，在高线公园建成后，废旧厂房的更新租赁率大幅度增加。以高线公园为核心展开了高端酒店、商品市集、创意工作室的新地产开发。高线公园挖掘了该片区的景观价值，并提升了切尔西

gardens primarily serves ornamental purposes, with limited consideration for public engagement. Therefore, even when gardens were open for visitation, they did not fully embody the concept of public parks. The world's first urban park emerged in Birkenhead Park, England. British architect Joseph Paxton conceived a park and investment property near Liverpool, England. The park's blueprint stipulated the allocation of 70 acres of public green space, intended to provide residents with recreational opportunities. The overall design incorporated residential structures encircling the park, affording unobstructed views of the verdant landscape. The park's layout simulated diverse topographical features, such as forests and lakes, to mimic natural scenery. The central promenade facilitated direct access to the commercial hub, while the secondary pedestrian ring remained uninterrupted, forming a circular leisure trail. The success of Birkenhead Park highlighted the significance of natural surroundings in enhancing real estate value, prompting urban planners to recognize the role of urban parks in regulating the urban environment and facilitating citizen activities. This realization sparked a global surge in urban park construction, with New York's Central Park standing as the most iconic and influential example. For over a century, the park has been designed to provide ecological preservation and cultural vibrancy to the city.

3.3.2 The value of the park

In this condensed historical account, we present the origins and evolution of urban parks, underscoring their paramount significance in shaping the future development of green resources in cities. The fundamental values inherent in a city can be categorized into three facets: environmental, economic, and humanistic values.

The environmental value pertains to the pivotal role played by the blue-green system in regulating the climate and ecology of urban areas. While city parks may appear as mere verdant spaces, they contribute to a multitude of natural ecological benefits that continuously protect the environment. For instance, the extensive establishment of parks and the strategic planting of indigenous trees along boulevards effectively mitigate the inundation pressure during the rainy season by absorbing large amounts amount of surface water. Moreover, the arboreal canopies

社区的文化艺术品位，为整个片区的商铺盈利、税收收入、就业率带来巨大的贡献。同时，因为高线公园的成功，纽约每年的旅游人数也直线攀升。同样，芝加哥市中心的千禧公园，也为本来废旧不兴的城市中心带来新的活力。

3. 公园与休闲绿地类型

19世纪开始，城市、郊区、乡村都掀起"公园和休闲"的公共服务产业热潮。小型规模的社区公园、游乐场、体育中心等娱乐活动中心开始分布在城市和乡村的居住场所附近。为了更好地规范公园设计，不同国家开始制定国家级公园、街区级公园、城市范围休闲娱乐设施的修建导则。例如，美国加利福尼亚州修建公园和娱乐设施的导则根据加州不同的气候区分而制定，在气候较为暖和的地区，人均拥有面积更大，也允许更大范围的树木种植。而

中国的《国家公园总体规划技术规范》（LY/T 3188-2020），是由中国国家林业和草原局发布的国家公园林业规范标准，其中总结了国家公园、自然保护区、森林公园、湿地公园及沙漠公园自然保护地的理论和实践规划，并制定了设计标准。例如，休憩项目应与自然景观和传统生产生活方式相协调，重点发展休闲项目与周边社区居民的关系。

在街区层级的范围内，公园和休闲区域可划分为街区公园、游乐场、休闲中心等。街区公园，指的是以小学为中心的社区服务范围的街区活动。游乐场主要为5—14岁的儿童提供室外的娱乐设施，同时为学龄前儿童提供至少0.8公顷的休闲景观区域。最理想的规划尺度为每15分钟的生活圈拥有1个街区游乐场。[58] 街区公园需要反映不同年龄层儿童的需求，并

and vegetative corridors formed by towering flora serve to temper the urban heat island effect. The concealed strata of soil, coupled with the presence of planted trees, offer a congenial habitat for the myriad insects, animals, and avifauna that inhabit the city, thereby fostering a stable ecological environment.

Residing and working in a good environment of urban parks enhance the economic value of proximate properties in an accelerated manner. These parks not only provide a natural and wholesome green environment but also offer ample space for recreational pursuits, daily repose, sports, and large-scale festivities. With the activation of urban vitality catalyzed by the presence of parks, the adjacent entertainment venues, eateries, and amenities flourish due to the large population. As exemplified by the resplendent transformation of the once derelict and sparsely populated Chelsea neighborhood subsequent to the completion of the High Line Park in New York, the rejuvenation of leased abandoned factories has been nothing short of remarkable. Concurrently, upscale hotels, commercial markets, and innovative studios have emerged as integral components of novel projects centered around the High Line Park. This park harnesses the intrinsic landscape value of the region, improving the cultural and artistic predilections of the Chelsea community, thereby offering immense value upon tax revenues, employment opportunities, and retail profits throughout the entire area. Simultaneously, the resounding success of the High Line Park has led to a significant surge in annual tourist influx to New York. Similarly, Millennium Park, akin to downtown Chicago, has breathed new life into a previously desolate city center.

3.3.3 Types of parks and leisure green space

Since the nineteenth century, the emergence of the public service sector including "parks and entertainment" has been witnessed in urban, suburban, and rural areas alike. Smaller-scale community parks, playgrounds, sports centers, and other recreational community centers have been established in close proximity to urban and rural settlements. In order to establish superior design standards for parklands, various countries have developed guidelines for national parks, neighborhood parks, and city-wide recreational facilities. For instance, in the case of California, the construction of

且考虑人车分流的空间形态，让老人和小孩都能较为便捷地从家里、学校直达公园。在美国的街区公园规划原则里，还提出社区的休闲娱乐设施包括游乐场、运动场、游泳池、俱乐部和社区活动中心。

城市范围的休闲活动区，可归纳为露天剧场、游艇俱乐部、滨水休闲绿地等，也可以是城市郊区的动物园、植物园、主题儿童乐园、露营绿地等。休闲娱乐场地的设计需求会随着市民生活习惯的变化而改变。在城市发展初期，拥有自然的绿化环境和湖泊水体就是一个较为完善的休闲场所。但随着现代城市生活品质的提高，市民对于休闲服务的要求也相应提高。例如，城市滨水空间往往需要满足户外运动爱好者的需求，上海的很多滨水空间设有自行车道、滑板设施、攀岩设施、沐浴饮水服务驿站等供运动爱好者使用。传统观赏性的植物园和动物园也常常增加种植和喂养的互动活动，以及夜间观赏的活动，为市民提供不同的休闲体验。

由此可见，城市的公园和休闲服务设施会随着人口、工作和休闲模式的变化而改变，就像一百年前的公园形态也在随着时代的变化，改变着它的功能和形态。对于设计师来说，不能一成不变地复制导则标准，而要以敏锐的触觉了解生活需求的变化，思考未来的休闲景观模式形态。

四、社区设计

本节内容主要通过介绍社区建设的过程，让学生了解社区与自然环境的关系，从而对社区环境的建构形成、规划布局、设计要素产生

parks and recreational facilities necessitates adaptation to the different climatic characteristics of the region. In warmer climates, the per capita area is expanded, thereby allowing for a greater number of trees to be planted.In China, *the Technical Regulations for the National Park Master Plan*, a national park forestry standard issued by the State Forestry Administration and China Grassland Bureau, encapsulates the theoretical and practical planning of nature reserves, while also formulating design standards for national parks, nature reserves, forest parks, wetland parks, and desert parks. In open-space projects, it is necessary to coordinate the natural landscape with traditional production and lifestyle, while also fostering a symbiotic relationship between leisure projects and the residents of the surrounding community.

At the neighborhood level, parks and entertainment areas can be categorized into neighborhood parks, amusement parks, and entertainment centers. Neighborhood parks pertain to community-centric activities in residential areas, with primary schools serving as focal points. Playgrounds primarily cater to outdoor recreational facilities for children aged 5 to 14, with a minimum of 2 acres of landscaped recreational area also being allocated for preschool-aged children. The optimal planning scale involves the provision of a neighborhood playground in a 15-minute walking distance from residential areas.[58] These neighborhood parks must address the needs of children across all age groups, while also considering the segregation of pedestrians and vehicles, thereby ensuring convenient access for the elderly and children traveling from their homes to neighborhood playgrounds. The principles outlined in the US Neighborhood Playground Plan further advocate for the inclusion of playgrounds, sports fields, swimming pools, clubs, and community centers in community recreational facilities.

The urban landscape comprises a variety of recreational spaces, including amphitheaters, yacht clubs, waterfront recreation greens, as well as suburban recreational establishments such as zoos, botanical gardens, themed children's parks, and camping greens. As societal habits evolve, so do the design requirements of these leisure destinations. In the early stages of urban development, a

更深刻的认知。

1. 社区与自然的关系

社区是人类群居生活而形成的空间形态。[59] 社区由住宅、交通、医疗服务设施、绿化服务设施、休闲服务设施、学校等共同组成。理想的社区是自然场地和景观环境的最佳组合。社区营造主要的考察点如下。

（1）了解场地生态和地质环境。住宅及社区需要建立在安全、具有良好生态自然环境的系统里，尽量以不打扰的方式，在考虑到地形、排水、植物、水源、动物等方面的情况下，营造共存的环境。可以结合场地的地形，了解地质承载力并分析开挖和坡地巩固的难易程度，选择区域地质稳定的地址进行建造。同时了解地质条件，分析土层的结构、蓄水量，以及地下水的流线，避免在开挖过程中造成水源

的污染和径流的破坏。充分利用地形的优势，如利用山谷或山脊营造视野开阔的建筑环境，打造风景宜人的社区。

（2）适应气候条件的变化。社区的营造需要通过居住建筑群来调整寒冷、闷热、干燥或潮湿的区域气候特征，从而在人居建造环境中调整社区微循环气候。依靠自然的条件，利用整片的绿植和高大乔木阻挡强风，建立树阵挡墙引导微风。从人居规划的角度，调整交通路线、市政道路、红线范围来配合建筑对气候条件的应对策略。

（3）减少负面环境作用。社区环境需要尽可能避免不良因素，包括人工环境带来的污染、自然灾害、视觉的障碍物等。如果无法调整选址的位置，则需要通过地形设计、生态修复、植被调整、场地隔离、视觉引导等方式避

natural green environment with lakes provided a relatively comprehensive leisure space. However, in contemporary times, as the quality of urban life improves, the public's demand for entertainment services becomes more notable. Take urban waterfront spaces, for instance, they have to cater to the needs of outdoor sports enthusiasts. In Shanghai, numerous waterfront areas are equipped with bicycle lanes, skateboarding facilities, rock climbing amenities, and bathing facilities to accommodate sports enthusiasts. Traditional ornamental botanical gardens and zoos have also incorporated interactive planting and feeding activities, as well as nocturnal viewing experiences, to offer the public different experiences.

Therefore, parks and leisure services in cities undergo transformations in response to changes in population dynamics, work patterns, and leisure preferences. Similar to parks of a century ago, their forms adapt to the zeitgeist. Designers must possess a keen sensitivity to the evolving ways in which we live and contemplate the future configuration of leisure landscapes, rather than merely replicating prescriptive guidelines.

3.4 Community construction
3.4.1 The relationship between community and nature
A community represents a spatial form of human social life[59], including housing, transportation, medical services, greening services, leisure services, schools, and other amenities. The ideal community embodies an optimal combination of natural surroundings and a landscaped environment. The following key considerations are pivotal in community development:

Understand the site ecology and geological environment
Firstly, habitats and communities should be established in secure and ecologically sustainable natural systems. This involves considering topography, drainage, vegetation, streams, and avian life. The objective is to create spaces that coexist harmoniously without causing disruption. By analyzing the site's topography, we can select a geologically stable foundation, taking into account load-bearing capacity and the ease of excavation and slope consolidation. Additionally, understanding the geological conditions pertaining to soil structure, water retention capacity, and groundwater flow is crucial to prevent water

免负面的影响。[60]

（4）适应本土景观特色。社区的环境选址须适应本土景观特点。只有适应当地的气候地形条件，进行合适的规划，才能形成和谐的居住环境。如果是反本地景观条件而行的建筑规划形式，将会付出较大的危机成本，不利于社区的稳定发展。

2. 社区的构成要素

（1）场地布局

社区是群体生活的庇护区。人类有史以来都是依赖自然环境营造适宜的居住地，从而形成乡村和城市。在美洲拓荒时期，围绕着海湾和码头，以及在交通路线交会处，形成了各类型的居住区。社区的场地布局与自然环境质量、交通可达性、建筑服务设施分布、治安等多方面的因素都有关系。

良好的社区规划必须先对资源（景观、人群、社区一天内的各种活动等）进行现场考察。好的规划绝不会始于强加于社区的抽象和独断的策划，而是始于对现存条件和机遇的认知。[61]规划社区的场地布局时，首先要了解场地周边环境是否具有危险或者不健康的元素，其中包括：被污染的土质环境、具有化学污染排放的工厂区、周边水源有污水废沟、靠近发电站及水库地区，以及具有大面积山泥坡地、洪水高发区、地震震中地区等，这些对于社区居住规划来说都不是良好的环境。[62]另外，不健康的居住因素不仅是环境带来的危害，还包括居住体验的干扰与不适。例如，居住在道路繁忙的高速公路要道附近，车辆早晚的嘈杂声也是影响社区居住体验的不健康因素。社区犯罪率高、空置建筑较多、街道卫生环境恶劣、街区

source contamination and runoff during excavation. Leveraging the advantages of the terrain, such as valleys or sections of ridges, allows for the creation of expansive vistas and aesthetically pleasing built community environments.

Adapt to changes in climatic conditions
Residential buildings necessitate consideration of the climatic characteristics prevalent in the area, including cold, hot, dry, or wet conditions. In accordance with the natural environment, extensive clusters of verdant and towering trees are strategically positioned to serve as barriers against forceful winds, while the arrangement of trees facilitates the guidance of gentle breezes. From an urban planning perspective, the design of traffic routes, municipal roads, and red lines revolves around the requirements of the buildings, contingent upon the prevailing climate conditions.

Reduce negative environmental effects
The community environment must strive to evade detrimental social factors and environmental pollution predicaments to the greatest extent possible. Adverse

factors include environmental pollution, natural calamities, and visual obstructions. In instances where the site's location cannot be adjusted, it becomes necessary to mitigate the negative impacts through topographical modifications, ecological restoration, vegetation adjustments, increased distance from the site, and aesthetic enhancements.[60]

Consider local landscape characteristics
developmental attributes of the local landscape. Only through planning that aligns with the local climate and terrain conditions can a harmonious and environmentally conscious living environment be forged. Architectural planning that contravenes the conditions of the local landscape incurs a greater crisis cost and hampers the stable development of the community.

3.4.2 Elements of community
Site layout
Historically, human beings relied upon the natural environment to establish habitable settlements, thereby giving rise to villages and towns. During the pioneering era

建筑衰败及房屋遭蓄意破坏等影响社会安全的因素，都会形成不健康的社区建造环境。

（2）建筑

建筑的布局是社区设计的重要环节。建筑朝向的选择，决定了室内的日照受光程度。朝向南面的建筑立面，受到日光的照射时间更长，会给起居室、寝室、教室、工作空间带来较好的自然光照，同时也能节省建筑的用电。再比如，建筑坐落在山脊还是坡底，也决定了建筑的通风效果等。

不同建筑类型的社区，布局也会有不同的设计侧重。对于工业和商业体社区，建筑群落以商业园区、办公区、研究实验室、轻工业制造、仓库和其他配套设施为主。交通可达性流线、停车空间，以及社区内的文娱服务配套设施为主要的规划考量因素。对于零售商业开发

社区，布局规划与周边邻里、社区和地区有紧密的联系。小尺度的零售商业中心依赖于居住社区邻里的步行可达性，大部分处于1千米的步行可达（约15分钟）的生活区域内。[63] 中型尺度的商业开发，需要考虑城市市内与周边的商务零售联系，顾客辐射范围约在3千米（约45分钟车程）。[64] 至于大型的购物中心和零售店，则需要考虑更大辐射范围内的贸易区顾客，约6千米（约1.5小时车程）的服务需求规划，在这种尺度的零售规划下，机动车交通系统的便捷性尤为重要。

住宅群落的社区布局，开发商设定的人口密度及公共空间的比例分配，是住宅开发需要面对的规划重点。人口密度大，建筑户型分布需要呈现高层建造模式或者分散型模式，所形成的公共空间需求也随之而变化。法国马赛公

of America, diverse forms of settlements emerged around bays, docks, and the intersections of major transportation routes. Therefore, the layout of a community is devised in accordance with the quality, accessibility, distribution, and safety of building services in the natural environment. "Good community planning must begin with the on-the-spot analysis of resources (the landscape, the people, the various activities in the community in one day). Good planning never begins with the abstract and arbitrary planning imposed on the community, but begins with the understanding of existing conditions and the opportunities."[61] In the context of site layout oriented towards the community, it is necessary to first determine the presence of hazardous or unhealthy elements in the surrounding environment of the site. These elements include contaminated soils, areas housing factories that discharge chemical substances, regions with sewage outfalls in proximity to the water supply system, areas in close proximity to power stations and reservoirs, as well as areas characterized by steep hillsides, flood-prone regions, and earthquake epicenters. Such areas do not provide conducive environments for planning community life.[62]

Building

The architectural layout assumes a pivotal role in community design. The orientation of the buildings is chosen to determine the extent of sunlight exposure in the indoor environment. Buildings facing south are exposed to daylight for a longer duration, thereby facilitating superior natural lighting in accordance with the layout requirements of living rooms, bedrooms, classrooms, and workspaces. Additionally, this approach contributes to energy conservation. For instance, the positioning of a building on a ridge or at the base of a slope determines the ventilation in the structure.

Different types of buildings have different design standards for weight measurement. In the case of industrial and commercial communities, building clusters are planned to include commercial parks, office areas, research laboratories, light industrial manufacturing facilities, warehouses, and other auxiliary amenities. The primary considerations in community planning revolve around ensuring accessibility, providing ample parking space, and offering recreational and cultural services within the community.

寓[65]是现代高层住宅的雏形（图3-08）。为了不占据更多的占地面积，马赛公寓的公共空间被设置为一层的休闲和停车空间，而用户则使用屋顶花园作为社交场所，从而减少建筑面积的开发，最大程度地使用建筑规划面积。另外，住宅社区的设计还需要考虑硬质透水面积的比例。住宅的开发和公共活动空间会大面积使用硬质铺装供住户通行，这时便需要使用可透水铺砖以减少地表径流的雨水，从而减少雨水对附近河流区域的污染。

城市中商业、工业和居住社区是紧密相连的一体，但各自形成的社区环境在规模、朝向、布局和开发密度上有不同的要求。设计师需要了解本土城市的自然条件和社区开发类型，并在此基础上进行设计。

（3）路网格局

社区的规划离不开交通路网的布局。良好的交通网络有利于社区活动的有效进行，并能保障车行和人行的安全。组织网络布局的类型如下。

网格化布局。方格网的交通布局有利于城市网络的快速建造和施工，多修建在平原上，并能形成明确的方向指示效应。网格化的社区布局以西班牙巴塞罗那（图3-09）、美国纽约（图3-10）的社区最为典型。网格化有利于步行交通，对于复杂车流有较强的协调性。但网格化的缺点在于不适用于地形特征较强的场地，起伏较大的山地如果需要建造笔直的路网形态，需要更多的成本进行地形调整。此外，网格化对于交通量大和小的道路情况的处理是无差别的。

In the context of retail business development communities, retail enterprises aim to establish close connections with neighboring blocks, communities, and regions. In the case of small-scale retail commercial centers, residential areas should be easily accessible by foot. It is recommended that the majority of retail commercial centers be situated within pedestrian-friendly residential areas within a 1-kilometer radius.[63] Medium-sized retail developments, on the other hand, need to take into account the commercial retail connectivity in and around the city, ensuring that customers can be reached within a 3-kilometer radius.[64]

For large shopping centers and retail stores, a broader customer radius in the trade zone needs to be considered, typically around 6 kilometers, in order to adequately plan for service demand.

Concerning population density and the distribution of public spaces, residential development must prioritize the planning objectives set by the developer. In instances of high population density, buildings are typically distributed in the form of high-rise models or decentralized models, resulting in corresponding changes in the requirements for public spaces. The Marseille apartment [65] complex in France serves as a quintessential example of a high-rise residential building (Fig 3-08). To minimize ground space occupation, public spaces are designated as entertainment and parking areas on the ground floor, while residents utilize the rooftop garden for social purposes, thereby optimizing land utilization. Additionally, the design of residential communities must also consider the proportion of hard permeable areas. Extensive hard pavement is utilized in residential developments and public activity spaces. To mitigate surface runoff and prevent pollutants from entering nearby river areas, these large hard-paved areas necessitate the use of permeable paving blocks. The interconnections among the business, retail, commerce, industry, and residential sectors in the city are closely interlinked. Nevertheless, the community environments that emerge from each sector exhibit variations in terms of scale, orientation, layout, and development density, contingent upon the different community types. Designers must possess an understanding

图 3-08　**Unité d' Habitation, France**
法国马赛公寓

图 3-09　**El Eixample, Barcelona**
巴塞罗那扩展区

图 3-10　**Manhattan, NYC**
纽约曼哈顿

放射状布局。放射状的交通布局以城市节点为主要放射中心，面向场地四周扩展。其中典型的放射状社区网络是法国巴黎的规划。放射状的道路对于地形布局的适应性较强。一中心、多交叉的网络布局极为便利地增加了区域之间的连通性。同时，指向社区焦点的道路方向也十分明确。网络化的街道布置有利于居民进行自由的步行体验，而放射状的网络布置会因为道路指向多样，造成交通效率低下。同时，交通路网划分的区域地块形状不一，为后续的建筑设计带来挑战。

枝状布局。树枝状线路[66]，犹如树枝的枝干，道路向不同方向发展延伸。枝状道路能够较好地适应自然要素和有限制条件的地形。另外，在设置道路等级的过程中，能明显区分主干道、次干道和支路，为不同交通量的道路提供容量差异化设计。枝状网络不断发展延伸的道路系统有利于周边潜在的社区开发建构。但枝状道路的方向感较差，并且在支路容易形成末端的中断。枝状道路网络是城市及社区普遍使用的网络方式，它的无限延伸对不同网络系统都有连接的可能性。这为适应不同时期的社区扩建和发展带来极大的便利。

线形网络布局。线形道路主要在两点之间进行连接。这种连接方式对铁路、河流、高速公路及其他线形要素的开发指出明确的道路方向。但线形道路也有明显的弊端，单一的线形支路缺乏社区中心，在交通线路繁忙的地区容易产生大量的拥挤和冲突。特别是机动车道和步行道的双向需求，为步行安全带来隐患。

3. 社区的问题与未来

社区是我们每天生活、居住、共享家庭美

of the inherent natural conditions that govern the development of the local city and its communities.

Road network pattern

Community planning cannot be separated from traffic network planning. A good transportation network is helpful in carrying out community activities and ensuring the safety of vehicles and pedestrian traffic. The organization network layout types are as follows.

• Gridding: The grid traffic layout fosters the expeditious establishment of urban networks, primarily on level terrains. The grid layout is exemplified by cities such as Barcelona and New York City (Fig 3-09, 3-10). The grid network facilitates pedestrian traffic and exhibits robust coordination in managing traffic flows. However, the drawback of the grid system lies in its incompatibility with sites characterized by rugged terrains. Constructing a linear road network necessitates additional time to accommodate the undulating mountainous topography. In addition, the grid layout fails to differentiate between roads with high traffic volume and those with low traffic volume.

• Radial: The radial traffic layout revolves around urban nodes serving as primary radial centers, with the network layout oriented towards the surrounding areas of the site. Paris exemplifies a typical radial road pattern. The radial road length suits well with terrain layout. A central, multi-crossing network layout greatly enhances connectivity between various areas. Moreover, the roads exhibit a clear directional focus towards the central region. The street layout affords residents the freedom to traverse the area on foot. However, the radial network layout, due to its changed directional orientation, can also impede traffic flow efficiency. Simultaneously, the diverse forms of land blocks resulting from the division by the traffic network pose challenges for subsequent architectural design.

• Dendritic: The branching network[66], akin to the branches of a majestic tree, extends its reach throughout the course of road development, diverging in various directions. Straight roads, possessing a superior adaptability to the natural elements, reflect a remarkable capacity to conform to topographical constraints. Moreover, the process of establishing road grades effectively demarcates primary, secondary, and feeder road grades, thereby affording differentiated design considerations for road capacities

好时光的场所。如何让我们的社区健康、持久地发展，并使社区环境品质得到提升，这是设计师需要为居民考虑的更为长远的问题。首先，社区的安全问题。随着社区人口的增长，不同年龄、性格、背景、职业的居民，都要建立社区环境安全和健康安全的一致意识。这种意识的培养脱离不了对社区环境品质的维护。如果社区总是脏乱不堪，设施破败缺乏维修，又或者楼道和基础设施经常被随意涂抹和破坏，那么便很难唤起大家保持社区美好的初心。其次，社区场景的单调问题。社区场景都是以功能性强的建筑为主，户外景观则以易于维护的低成本绿化种植为主。对于常规化的绿化场景，物业没有充足的预算经常进行绿植维护和季节性花卉置换，社区景观的多样性展示会非常有限。社区可考虑组建自发的管理团队，以居民参与的方式开展花卉或者蔬果的种植，不仅可以降低维护团队的成本，也能拉近居民与社区的关系。同时，使用本土植物进行种植的配置和规划，也能让社区保持较为多样的景观体验层次。最后，社区的容量问题。每个项目的开发都有一定的人数容量，但对于老旧小区和历史性城镇来说，现有的社区空间结构已经不能满足现代居民的使用需求。面对不能扩容的社区环境，人口数量的增长、年龄层次和使用习惯的不同、公共空间配分不均等问题，是社区的更新改造面临的巨大挑战。在进行社区的更新设计时，设计师需要细致地梳理现有的街区和交通流线，理解住户之间的居住需求，通过公共活动空间促进社区和人的和谐共处，把设计的创意转化为具体的设计策略，让居民通过环境的改变感受到真正的"社区"。

commensurate with varying traffic volumes. The network of branched roads creates a comprehensive system that can be continuously developed and extended to include potential communities in the surrounding vicinity. However, the inherent nature of branch roads, characterized by their lack of regular direction, tends to give rise to terminations at the junctures of these branches. The branched road network, a prevalent network typology employed by cities and communities, features boundless possibilities for expansion and the potential to interconnect diverse network systems. Therefore, it offers unparalleled convenience in accommodating the growth and development of communities at different temporal junctures.

• Linear Networks: Linear roads primarily connect two points. This mode of connection provides a clear direction for the development of railways, rivers, highways, and other linear elements. However, linear roads also have evident drawbacks. Singular linear branches lack community centers, leading to potential congestion and conflicts, especially in areas with busy traffic routes. This configuration, particularly with the bidirectional demands of motor vehicles and pedestrian pathways, poses safety concerns for pedestrians.

3.4.3 The challenges and the future of the community

The community, as a dwelling place where we reside and partake in the joys of familial existence on a daily basis, necessitates a concerted effort to foster sustainability and enhance its environmental quality. This longer-term challenge demands thoughtful consideration of designers in their quest to improve the well-being of residents.

Foremost among the concerns is the security of the community. As the population burgeons, individuals of all ages, backgrounds, and occupations must possess a shared understanding of the community's safety and health. Without the ability to uphold the quality of the community environment, the cultivation of such awareness becomes an arduous task. Should the community's green spaces consistently languish in a state of disarray, its facilities fall into disrepair and suffer from neglect, or its corridors and infrastructure become subject to haphazard and wanton acts of vandalism, the desire to preserve the community's aesthetic attraction becomes increasingly elusive.

社区的规划对人类的健康存在影响吗？我们可以把影响因素分为自然因素和人为因素。自然因素方面，我们可以通过选址、建筑设计、场地修复的方式让社区成为安全的区域。可是在人为因素上，规划设计如何引导居民在使用的周期中让社区形成健康有活力的核心区呢？在未来的社区规划中，设计师需要有意识地营造居民健康生活的轨迹，这种健康轨迹可分布在公共广场、景观绿化、娱乐设施，以及体育活动场景中。通过优质美好的景观场景和社会活动，激发人们对于社区环境的自治管理和爱护，从而打造健康的社区。

小结

从城市空间的发展历史，可以观察到城市形态的形成与时代的功能需求有着密切的关系。城市形态组成的主要元素和特征也随着时间而发生变化。学生需要在日常体验中对环境的变化保持敏锐的触觉，不断地思考构成问题的原因和解决方法，不断地向自然环境学习。

Secondly, the issue at hand pertains to the monotonous nature of the community setting. The residential area's landscape predominantly comprises utilitarian structures, while the outdoor environment primarily features cost-effective greenery that requires minimal upkeep. Due to budgetary constraints, the hotel lacks the means to regularly maintain and replace seasonal flowers in the traditional green landscape, resulting in a limited diversity of community scenery. To address this, communities may contemplate the establishment of self-directed management teams, wherein residents actively engage in planting and cultivating fruits and vegetables. This approach not only reduces landscape maintenance costs but also fosters a closer bond between residents and the community. In addition, the utilization of indigenous flora will enhance the park's landscape, providing a more multifaceted experience.

Finally, the issue of community capacity arises. Each development project possesses a finite capacity for growth. However, in the case of ageing blocks and historic towns, the existing spatial structure can no longer adequately cater to the needs of modern residents. Confronted with an unexpandable community environment, an expanding population, varying habits among different age groups, and an imbalanced utilization of public spaces, community regeneration faces significant challenges. In the process of community revitalization, designers must assess the existing blocks and traffic flow, understand the residents' requirements, and strive for harmonious coexistence between the community and its inhabitants through the utilization of public spaces. By translating design concepts into effective strategies, residents can tangibly experience the transformation of their environment, thereby fostering a genuine sense of community.

Summary
The factors influencing community development can be categorized into natural and human factors. From a natural standpoint, site selection, building design, and site restoration can transform the community into a planned and secure area. However, when considering human factors, how can planning and design guide residents towards cultivating a healthy and vibrant core through the cyclical utilization of communal spaces? In future community planning, designers must consciously forge a path towards a wholesome lifestyle for residents. This trajectory can include public squares, landscaping, recreational facilities, and physical activity settings. By means of high-quality landscapes and social activities, individuals will be motivated to independently manage and protect their community environment, thereby creating a robust framework for healthy community planning.

课后作业
After-class exercises

讨论题
Discussion

如何界定场地设计的优势和劣势？请结合校园中的具体案例进行讨论。

Please consider the advantages and disadvantages of the site design of the campus you are most familiar with? Please discuss with specific details.

设计实践
Design practice

"最好的场地规划是以最小的总成本和阻力取得最大的长期效益"，请结合课程内容，选择相关的场地规划案例，分析其对于长期发展的规划价值和设计方法，并用图纸的形式进行分析表达。

"The best site planning is to get the maximum long-term benefits with the minimum total cost and resistance." In the context of the course, please select relevant examples of site planning and analyze planning values for long-term development from the perspective of design methods.

推荐读物
Reference reading

1. Cuff, D., et al. *Urban humanities: New practices for reimagining the city*. MIT Press, 2020.
2. Di Mari, A. and Yoo, N. "Operative design." *Amsterdam: BIS*, 2013.
3. Foster, S. R. and Iaione, C. *Co-Cities: Innovative Transitions toward Just and Self-Sustaining Communities*. The MIT Press, 2022.
4. Giedion, S. *Space, time and architecture: the growth of a new tradition*. Harvard University Press, 2009.
5. Goh, K., et al. *Just Urban Design: The Struggle for a Public City*. The MIT Press, 2022.
6. Lewis, P., Tsutumaki, M. and Lewis, D. J. *Manual of section*. Chronicle Books, 2016.
7. Litscher, M. "Jane Jacobs: the death and life of great American cities." *Schlüsselwerke der Stadtforschung*, 2017: 367-394.

第四章

人性化景观设计

Chapter 4

Human-centered landscape design

学习要点
Study Points
人性化景观设计的原理和方法
Principles and methods of human-centered design

学习目的
Teaching Objectives
学生通过案例分析和研究视角观察，拓展人性化空间设计的新思路。
Enable students to develop new ideas for the design of human-centered spaces through the analysis of case studies and the observation of research perspectives

这一章主要探讨景观设计中人性化场景的应用方式。章节从景观环境与人性化设计的关系、人性化景观场所的设计、无障碍设计、人性化的服务环境四个主题进行人性化景观专题的讨论。通过上述介绍，让学生了解人性化设计的原则、方法，以及在实际景观空间中的应用。

This chapter primarily discusses the application of human-centered scene design in landscape design. It is divided into four categories of human-centered landscape themes: landscape environment and behavioral characteristics, types of human-centered landscape sites, principles of barrier-free and universal design, and environmental service design. By introducing these subsections, the reader is acquainted with the principles of human-centered design, design methodologies, and the practical application of design in landscape spaces.

一、景观环境与人性化设计的关系

人居环境发展过程中，既要尊重自然场地的生态变化，又要照顾到居民不同的年龄、性别、健康程度，从而调整环境的设计。本节从人性化设计的定义、物理层次、心理层次、社会层次四个方面进行介绍。

1. 人性化设计的定义

人性化设计，旨在符合人的行为习惯、生理结构、心理需求、社交需求，以及思维特性的设计功能和活动策划。[67] 而人性化场所，在本章节以城市的公共空间为主要讨论对象。以使用者在公共场所的舒适度、便利性、愉悦度等体验对城市公共空间进行人性化设计考量，体现出城市人文关怀的温度，也展现出对人性的尊重。[68]

当我们说起一个城市的时候，它的气质和文化特征往往与其生活场景联系在一起。城市的公共空间充满了人类生活的不同场景。中世纪的城市广场，是人们聚集议事、庆祝节日、宣讲理念、执行判决的地方；同时也是市民采买、打水、社交、讨论时事的场所。中国古代"天人合一"的理念，反映了对人的行为、精神、思想的尊重。历史上，对人性的再发现可以引起时代的变革。欧洲文艺复兴运动起源于艺术家、文学家以及科学家对于人为本体的科学认知以及人性的解放。人的美、自由的理念、自由的言论是被鼓励的，从而产生了划时代的艺术作品和文坛著作。达·芬奇、拉斐尔、米开朗琪罗等文艺复兴艺术大师都是该时代的领军人物。

在当代的城市背景下，公共空间是展示城市政治和文化核心的重要场景，也是城市生活

4.1 Landscape setting and human-centered design

In the process of human settlement development, it is essential to respect the ecological development of natural sites while also addressing the adjustments in the environment's design to accommodate changes in age, gender, and health. This section introduces the concept of human-centered design through discussions on its definition and physical, psychological, and social attributes.

4.1.1 Definition of human-centered places

Human-centered design aims to plan and design functions and activities based on the characteristics of human behavior, physiological structure, psychological needs, social needs, and cognitive processes.[67] However, this section focuses on humanizing urban public spaces, considering the comfort, convenience, and enjoyment that individuals experience in these areas. Human-centered public space design serves as a means to reflect the vitality of the city and demonstrate reverence for human nature.[68]

When discussing a city, its spiritual temperament and cultural attributes are often intertwined with its vibrant scenes. During the Middle Ages, town squares served as gathering places for meetings, celebrations, religious proclamations, and even acts of punishment. They also functioned as venues for citizens to receive daily news, purchase food, fetch water, engage in discussions on current affairs, and socialize. The public spaces of the city were teeming with diverse scenes depicting the active lives of its inhabitants. In ancient China, the concept of "harmony between man and nature" reflected the perspective of ancient philosophy. This viewpoint also exemplified the reverence ancient Chinese thinkers held for human behavior, spirit, and cognition. Western societies also underwent a transformation in their appreciation of human nature. The Renaissance, which emerged in the fourteenth century, originated from a scientific understanding of human nature. It liberated human potential through artists, literary scholars, and scientists, promoting the celebration of natural beauty, individual freedom, and freedom of expression. This movement gave rise to groundbreaking artistic and literary works. Renaissance masters such as Leonardo da Vinci, Raphael, and Michelangelo were prominent artists figures in that era.

活力和乐趣所在。例如，罗马凝聚了艺术和追求美好生活的精神，在各大广场都集聚了历史性的雕塑作品，琳琅满目的市集、公共的座位台阶供大家享受地中海的阳光和生活的气息。在北欧丹麦，城市街道取代了忙碌的高速公路。非机动车的交通场景使丹麦成为市民生活幸福感最高的城市之一。其中，对人性化设计的考虑渗透了城市的各个空间细节。形态各异的自行车、残疾人步道以及能辅助母婴车出行的公交车系统，都是丹麦城市公共空间的重要关注点。[69] 城市的公共空间已经与城市的功能、形态、人居生活模式形成相互作用。人性化的设计模式，改变着城市的生活气质和人文精神。

景观设计中的人性化设计，不仅体现为以用户需求为核心对空间场所的优化，更体现为对地区文化精神的提炼。具体体现在户外场地的空间营造上，功能类型主要集中为城市广场、邻里公园、小型口袋公园、大学户外公共空间、住宅户外空间、儿童户外空间、医疗户外空间等城市空间形态。

2. 景观环境的物理层次关怀

景观环境在物理层次的人性化关怀，可以理解为功能需求的合理性。主要体现在景观空间场所的功能配置、自然资源协调的合理性以及体验过程中是否会得到安全舒适的建筑环境保护。在景观场景规划中，日照、风力、遮阴等因素在不同季节都要得到合理的功能安排，从而为使用者的日常生活提供关怀。例如，在城市广场中使用大面积的防滑和透水性较强的材料，可在雨水天气给步行者提供保护。防滑的铺装以及排水设计较好的广场，在雨季来临时可以快速缓解雨水聚积的问题，避免形成积

In the backdrop of contemporary urban landscapes, public spaces assume a pivotal role in showcasing the cultural and political essence of a city, as well as the vibrancy and exuberance of urban life. For instance, the city of Rome epitomizes the spirit of art and the pursuit of an enhanced existence. Historical sculptures grace the main city squares, markets, and public seating platforms in the Mediterranean, affording everyone the opportunity to bask in the warmth of the sun and revel in the joys of life. In contrast to bustling motorways, the streets of Denmark, a nation renowned for its profound appreciation of urban well-being, present a scene devoid of motorized traffic. This is a testament to Denmark's commitment to human-centric design, which permeates the spatial intricacies of each city. A diverse array of bicycles, thoughtfully designed sidewalks for the disabled, and a comprehensive bus system catering to the needs of mothers and infants constitute integral components of Danish public spaces.[69] Therefore, the urban fabric has evolved to align with the city's living patterns, as the human-centric design paradigm reshapes the residential temperament and humanistic ethos of the urban landscape.

Humanistic design in landscape architecture not only underscores the optimization of spaces with human needs at its core, but also embodies the refinement of regional culture, primarily through the creation of outdoor spaces. The functional typologies of such design are predominantly concentrated in urban spatial configurations, including city squares, neighborhood parks, small pocket parks, outdoor public spaces, residential outdoor spaces, outdoor play areas for children, and outdoor spaces for medical facilities.

4.1.2 Physical level care of landscape environment

Humanistic care of the landscape environment at the physical level can be understood as the rationality of functional requirements. This is mainly reflected in the allocation of land to function, the coordination of natural resources, and whether the built environment is safe and comfortable for its intended purpose. During the landscape planning process, factors such as sunlight, wind, and shade must be meticulously considered across different seasons, and judicious functional arrangements must be devised to cater to the exigencies of daily life. For instance, the utilization of extensive anti-slip and permeable materials

水造成滑倒。在功能布局上，建筑入口会提供遮阳棚和缓冲停留区，同样也是为满足使用者的人性化空间需求。遮阳棚不仅在夏季提供出入口的遮阴功能，在突发大雨和大雪天气时也能为等候车辆的人群提供暂避的地方。在处理出入口高差问题的细节方面，以弧形的路沿石来巧妙处理高差冲突，可以让轮椅和儿童推车都能较好地通行（图4-01）。

另外，材料的运用也使人性化设计的保护更为全面。例如，智能户外材料技术能很好地识别天气条件，在不同的使用场景中进行切换。智能化座椅表面温度和湿度的传感器能操控座椅进行加热或降温。在温度寒冷的雨雪天气，可以在户外提供温热效果，给使用者带来干燥温暖的休息体验；同样，在炎热暴晒的环境下，座椅可根据户外环境的变化进行降温，为使用者提供清凉的休息体验。智能化场景已经应用到户外活动的景观设计中[70]，如感应灯具、可显示公里数的跑道、可显示使用者运动指数的标识牌等，都为人性化的户外体验提供着重要的设计支持。

3. 景观环境的心理层次关怀

景观环境营造中的心理关怀主要侧重于建立安全、有教育意义的户外活动场所。[71]这给居民提供了缓解城市压力的活动机会，有利于使用者身体的健康和情绪的安宁。一方面，需要注意使用过程中不同人群需求的划分。例如，青少年竞技活动的场地，应该与老人、儿童以及特殊人群的活动场地有一定的区分，不能互相干扰。竞技类的场地，如足球场、篮球场、滑板公园、自行车道等，活动流线较为复杂，易发生肢体碰撞或者摔伤。为避免动态流线和

in city squares serves to safeguard pedestrians during inclement weather. Non-slip paving and well-designed drainage systems swiftly mitigate the issue of rainwater accumulation during the wet season, thereby averting the formation of stagnant water that may lead to slips and falls. In terms of functional layout, the presence of canopies and buffer zones at building entrances fosters a human-centric spatial experience. These canopies not only provide shade during the summer months but also offer shelter to individuals awaiting transportation in the event of sudden heavy rain or snowfall. Moreover, curved curbs facilitate the ingress of wheelchairs and strollers in scenarios where there appears to be a difference in height between entrances and exits (Fig 4-01).The selection of materials in the landscape confers a more comprehensive assurance. For instance, outdoor equipment technology is capable of identifying meteorological conditions and transitioning between diverse functionalities. In the case of seating arrangements, sensors for surface temperature and humidity can indicate whether the seat is heated or cooled. During inclement weather characterized by cold, rain, or snow, outdoor equipment can provide warmth, thereby affording users a dry and cozy respite. Similarly, in hot and sunny conditions, the seat can proffer a cooling function in response to outdoor activities. Concerning landscape infrastructure, technology has been implemented in the landscape design of outdoor activities.[70] For instance, the utilization of luminaires, thoroughfares incorporating motion indicators, and the relationship of signage with information pertaining to users' well-being all provide important additions to a human-centric outdoor scene experience.

4.1.3 Psychological level care of landscape environment

Psychological care in the creation of the landscape environment focuses on the creation of safe, educational places for outdoor activities.[71] This provides opportunities for residents to relieve the pressure of the city and contributes to their physical health and emotional well-being. It is necessary to pay attention to the needs of different groups of people. For example, the space for competitive youth activities should be separate from the space for activities for the elderly, children's activities, and activities for special groups, so that they do not interfere with each other. For competitive venues, such

Hotel entrance

图 4-01　酒店入口

as football fields, basketball courts, skating rinks, and bicycle lanes, the activities are more complicated, and more prone to physical collision or falls. To avoid conflicts between dynamic and static flow, we can combine the areas for the elderly and children, away from walking areas, and avoid interfering with their activities. On the other hand, the daily maintenance of the public space can be strengthened, thus improving environmental safety. Children's sandboxes, interactive fountains, sculptures, and other activities can encourage residents to spend time in the outdoor environment, and increase their security and concern for the site. When people participate in the process of designing, creating, and maintaining a public space, they take pride in the public space as individuals or groups. This bottom-up mode of public governance allows residents to "privatize" public space. This kind of privatization is a way to keep public space in the residents' consciousness as if it were their own home.[72] Through activities and management autonomy, people are psychologically connected with public spaces, which is a pleasant, comfortable, and meaningful place in daily life, thus becoming a place for people.

4.1.4 Social level care of landscape environment

The significance of the accessibility of pedestrian environments surpasses mere aesthetic appeal, as asserted by psychotherapist Joanna Poole. She posits that indulging in leisurely activities at outdoor cafés not only provides enjoyment but also plays a vital role in promoting the health and safety of urban areas.[73] The perception of safety and well-being in cities is connected to the availability of public spaces. Conversely, feelings of fear and danger are often derived from anxiety surrounding these communal area. While many t residential communities can be gated communities, but there are still a large number of adjacent buildings, suburban apartments and older blocks lacking all-weather security measure. Enclosed spaces, such as viaducts, abandoned structures, and subterranean pedestrian environments, devoid of pedestrian activity, can instill a sense of fear regarding personal safety. Therefore, it is necessary to consider the public's demand for safety in outdoor spaces. To minimize the rise of crime in urban settings, designers must effectively address three key challenges during the urban planning process.

静态流线的冲突，我们可以将老人、儿童活动区域和漫步区域的功能结合在一起，避免老人和儿童在活动的过程中受到干扰。我们还可以加强公共空间的日常维护和联系，增强社区对于环境安全的监管。户外环境中可加入参与性或者灵活布置的种植花池、儿童沙池、互动喷水池、雕塑等，增强居民与场地的互动，由此提升场地的安全性和关注度。在公共空间的设计、营造及维护的过程中，居民应该以个人或者团体的方式对该空间进行监管。这种自下而上的公共治理方式，让居民把公共空间视为"私有化"的场所，将公共空间当作自己的家一般去维护。[72] 通过互动活动以及管理自治，让人们从心理上将公共空间与日常生活中愉悦、舒适、富有意义的场景联结在一起，从而形成人性化的设计。

4. 景观环境的社会层次关怀

步行环境安全的重要性远比美学的吸引力强，心理治疗医生乔安娜·普尔认为在户外咖啡厅享受的时光不仅是愉快的消遣，更是城市健康和安全的必要元素。[73] 城市的安全感和幸福感，很多时候是公共空间直接带来的。同样，恐惧感和危险意识很多时候也来源于具有犯罪伤害性的公共空间环境。现有居住社区多以封闭式为主，但仍有大量的临界建筑、郊区公寓、老旧社区等缺乏全天候的保安看护。又或是高架桥下、废弃建筑群，以及地下人行道这种较为封闭又缺少行人参与的空间，很容易引发场地不安全感。这些都应当引起我们对大众户外空间安全性需求的思考。

为避免城市犯罪环境的滋生，设计师需要在城市规划的过程中，有效解决以下三类问题。

Firstly, the planning of outdoor spaces necessitates a symbiotic relationship with community management. The coexistence of outdoor spaces and architectural structures can impede the maintenance of public security and community order. A prime example of transforming a neighborhood to foster a sense of camaraderie among residents is represented by the Topotek 1+BIG Architects+Super flex in Denmark[74] (Fig 4-02, 4-03). Prior to the design of the linear park, the area was an unadorned square situated in an economically depressed and disorderly immigrant community, akin to certain undesignated city streets. While individuals from diverse cultural backgrounds would socialize outdoors, their interactions were limited. The chaotic nature of the community also had an adverse impact on neighborhood safety. The designers introduced vibrant colors to highlight outdoor activity spaces, enabling neighbors to easily identify areas conducive to interaction and joint exercise. Drawing inspiration from sports facilities catering to various cultural groups, the design integrated recreational amenities, thereby fostering a sense of belonging among community residents. Notably, this public space comprises

national infrastructures such as Muay Thai arenas, Indian rock-climbing playgrounds, Brazilian benches, classic British cast-iron garbage bins, and bicycle parking lots reminiscent of those found in Iran and Switzerland. The opening of the linear park sparked immense interest in the community. Residents from diverse cultural backgrounds congregate in this park to engage in cultural exchange, introduce their customs, and share their experiences. A once-avoided public space has been deftly transformed into a conduit for deepening community culture.

Secondly, the increase of recreational facilities in economically disadvantaged areas improves communal security. As expounded in *Healing for the City*[75], the increase of entertainment facilities exhibits a direct correlation with the diminution of criminality. In the blueprinting of novel communities, aside from the provision of novel residential abodes and transportation infrastructure, it is necessary to offer social intervention measures to enable the engagement of young individuals in athletic pursuits, thereby enhancing their well-being, while simultaneously affording women and children the opportunity to partake

第一，户外空间规划与社区管理形成紧密合作。户外空间和建筑空间如果像互不干涉的两个群体，那么公共环境的安全和社区环境的秩序就很难保持。丹麦的超级线性公园[74]（图4-02、图4-03）是典型的通过改造社区环境提升居民之间以及民族之间亲密感的例子。超级线性公园在改造前只是杂乱的移民区空置广场，如同众多缺乏设计的城市街道，不同民族的居民各自为政，没有过多的交集。而杂乱的社区人群也影响了社区环境的安全。设计师通过亮丽的颜色把户外活动区域展示给街区，让附近的居民能快速找到可以一起交流和运动的场所。同时，以各民族特色的运动设施作为设计灵感，将游玩设施与民族文化融合，从而增强人们对社区场所的归属感。如印度操场、巴西长椅、英国经典的铁铸垃圾桶、伊朗和瑞士的自行车停车位等具有各种民族特色的基础设施都可以在这个公共空间找到。线性公园的开放得到了社区居民的高度关注，各民族的居民在这个公园里共同交流、介绍、分享各自的文化。一个曾经引起社会不安的公共场地被设计师巧妙地转化为连接社区的纽带。第二，通过低收入地区娱乐场地的增加，来提高社区治安的意识。《治疗我们的城市》[75]中提到，娱乐设施的增加对犯罪率减少有直接的影响。在新建的社区规划中，除了新建住宅和交通基础设施，更重要的是提供社会干预，让年轻人可以通过运动场地增强体魄，妇女和小孩可以通过户外活动增加社区中的亲密度。同时，优质的社区环境还会提高建筑的入住率，从而减轻建筑空置率的问题。第三，增强公众决策参与，让户外空间无犯罪死角。科学而有经验的设计规划可以

in communal open-air activities. Concurrently, the establishment of a superlative communal environment increases the occupancy rate of buildings.

Thirdly, the fortification of public involvement in decision-making processes serves to curtail criminality in open-air spaces. The emergence of open-air spaces susceptible to perilous activities can be forestalled or circumvented through scientific and seasoned design planning.[76] This is most efficaciously and comprehensively achieved via community-led surveillance. Unimpeded public spaces, wherein denizens and passers-by maintain an unobstructed line of sight, foster perpetual connectivity. However, planning is not omnipotent, and those residents who have resided in the community for several decades possess a superior understanding of the environment vis-à-vis the designers. It is reasonable to engage the public in the course of community planning and participation, thereby formulating solutions tailored to local exigencies, rather than imposing preconceived designs. For instance, in the context of sight-line monitoring, the majority of activity areas are demarcated by hedgerows. Nevertheless, in the case of children and elderly individuals utilizing wheelchairs, hedgerows fail to offer genuine monitoring protection, but rather constitute an impediment. Therefore, designers must engage in discourse and experimentation with the residents, to proffer them with more pragmatic safety solutions.

4.2 Human-centered landscape

The planning of urban landscape spaces is categorized according to functional typologies: urban squares, neighborhood parks, city parks, street spaces, and thematic outdoor spaces. Design criteria are posited from three perspectives: the exigencies of the natural environment, social interaction, and infrastructure design.

4.2.1 Human-centered parks and squares

Parks serve as urban oases, catering not only to recreational requisites but also serving as pivotal ecological coordination sites in the concrete desert that is the city.[77] Parks afford denizens the opportunity to visually experience the changes of the four seasons, whilst concurrently providing a stable natural habitat for the fauna and flora inhabiting the urban landscape, and serving as serene and tranquil sanctuaries for

预防或者避免形成易引起不安全活动的户外场所。[76] 其中社区自身的监测是最为全面、有效的措施。在无遮挡的公共空间，居民和路人可通过视线对场地进行监测，随时发现危险信号。同时，设计师并非无所不能，生活在社区超过几十年的居民往往比设计师更了解社区的环境。公众重视和参与社区规划，形成因地制宜的解决方法，往往比设计师的方案来得更为合理。如刚刚提到的视线监测问题，大部分活动场所会使用绿篱对活动区域进行划分，可是对于儿童和需要使用轮椅的老人的视线高度来说，使用绿篱不仅没有起到保护监管的作用，反而形成了障碍。所以，设计师在完成设计方案后，需要与居民进行多次讨论和检验，从而在安全方面为社区居民提供更实际的保护。

二、人性化景观场所的设计

城市景观空间的规划，以功能类型为区分，主要以城市广场、邻里公园、城市公园、街道空间以及主题户外活动空间为主要对象进行设计分析。本节通过自然环境需求、社交需求、基础设施设计需求三个角度进行人性化设计标准的介绍。

1. 人性化的公园及广场

公园是城市中的绿洲，不仅满足活动休闲的需求，也是城市"混凝土沙漠"中重要的生态协调场地。[77] 公园为居民提供四季变化的视觉体验，为城市中的动植物和昆虫提供稳定的自然栖息空间，也为繁忙的城市居民提供宁静平和的精神场地（图4-04）。

既然公园是城市中人类活动最频繁的场所之一，那么公园场所人性化的理念对于设计规

the harried urban populace (Fig 4-04). Considering that parks represent one of the most frequented locations for human activities in the city, the concept of human-centric spaces in parks assumes a paramount role in design and planning. For the elderly, children, and individuals with mobility impairments, the humanization of park spaces constitutes a pivotal indicator of urban solicitude.

Design requirements of natural environments
The design requirements that should be paid attention to are as follows.

Firstly, the urban parks exhibit the highest density of plant coverage in the city, thereby underscoring the significance of preserving a diverse array of plant species to create a visually captivating landscape. By employing plant design, an integration of fruits characterized by different colors, shapes, fragrances, and botanical classifications is achieved. The varying heights of trees, shrubs, and vegetation further attract birds, butterflies, and insects, thereby forming a miniature ecological system.

Secondly, rivers, lakes, and oceans constitute integral components of the natural panorama. By integrating aquatic elements into the urban landscape through the inclusion of fountains, waterfalls, and artificial lakes, a serene and organic ambiance can be effectively conveyed through both auditory and visual means.

Thirdly, the presence of large trees forms a natural shade network that reduces ground-level heat, alleviating the urban heat-island phenomenon, thereby affording individuals shaded and comfortable pedestrian spaces (Fig 4-05). The sheltered environment, replete with seasonally regulated trees, facilitates the enjoyment of a myriad of outdoor activities by the public (Fig 4-06, 4-07). The combination of trees of varying statures during the planting process creates social spaces characterized by spatial differences, while simultaneously fostering an appreciation for the cyclical transformations of the seasons. Whether it be engaging in shaded picnics, partaking in football matches, attending open-air music festivals, or indulging in private or familial reading sessions, all such activities take place in a natural environment, thereby ensuring that

图 4-02 Topotek 1+BIG Architects+Super flex in Denmark – The Black Square
丹麦的超级线性公园——黑场

图 4-03 Topotek 1+BIG Architects+Super flex in Denmark – The Red Square
丹麦的超级线性公园——红场

图 4-04 Green park
绿色公园

划就极其重要。特别是对老人、儿童，以及行动不便的人群来说，在公园休闲场地得到人性化的服务是城市关怀温度的重要指标。其中需要关注的设计需求如下。

（1）自然环境的设计需求。第一，丰富的植物营造。公园是城市中植物覆盖密度最大的区域之一，保持植物品种的多样性是营造丰富景观的重要部分。通过植物营造，把不同颜色、形状、香味和果实的植物进行组合，利用高低不同的乔灌木以及植被的变化吸引鸟类和昆虫，形成小型的生态循环圈。第二，水体部分的营造。河流、湖泊和海洋都是自然景观重要的组成元素。将水资源元素融入城市景观，形成喷泉、瀑布、人工湖泊等，可以通过听觉和视觉的感受传递平静自然的气氛。第三，提供树下遮阴空间。大型乔木形成天然的遮阳网

络，能够降低地面的热度，缓解城市热岛效应，给予人们阴凉舒适的步行空间（图 4-05）。不同高低变化的乔木组合可以形成空间不一的社交场所，给予人们不同季节维度的空间感受（图 4-06、4-07）。具有季节性调节功能的树荫庇护环境，能够让市民享受各种类型的户外活动。在树荫下野餐、踢足球、演奏音乐，又或者享受私人阅读时光——自然环境提供的天然活动场所，无论炎热还是寒冷，正午还是晚上，都可以让人们舒适地放松休闲。

（2）对社交环境的设计需求。虽然城市公园和广场有着优厚的自然条件，亲近自然是人的天性，但是研究发现，大部分人去公园并不是因为景色优美，而是出于社交需求。[78] 因此，公园和广场所形成的空间形态很大程度上会对人们的社交方式产生导向作用。在这类空

individuals can locate a tranquil haven irrespective of the temperature or the time of day.

Design requirements for social environments

Urban parks and squares, endowed with unique advantages, inherently beckon humanity to commune with nature. However, Most of the research has revealed that the primary impetus for individuals frequenting parks is not solely the picturesque vistas, but rather the desire to accompany friends or relatives.[78] Therefore, the spatial configurations of parks and squares can lead to diverse social patterns of utilization. A human-centric design approach necessitates the provision of different social environments replete with divergent pathways, thematically designated activity zones, communal open spaces, and secluded social enclaves, all tailored to cater to the multifarious needs of the visiting populace.

First of all, a variety of exploration routes is provided, offering pedestrians diverse experiential journeys, while visually diverse scenes foster a shared thematic backdrop for social engagement. The ample width of the sidewalk accommodates uninterrupted pedestrian passage, ensuring minimal disturbance to individuals seeking respite.

Secondly, the inclusion of al fresco dining areas enhances social interactions. Outdoor dining and refreshment zones serve as pivotal gathering spots for families and companions. In parks and plazas, communal picnics forge stronger familial and group bonds. Site designs can incorporate outdoor furniture conducive to effortless dining or barbecuing, as well as public lawns tailored for picnicking, harmoniously blending with the natural surroundings.

Thirdly, the provision of spaces conducive to dialogue and communication is paramount. Apart from familial or friendly rendezvous in parks and squares, certain individuals seek solace in public spaces for personal relaxation. This demographic comprises recently employed youth, sun-seeking seniors, grocery-laden adults, and coffee-seeking workers. While these individuals may not necessarily require designated resting areas, their primary objective is to wander leisurely. To cater to this segment, designers must hold an understanding and attentiveness to their unique needs.

图 4-05　**Street trees**
行道树

图 4-06　**Metasequoia grove in winter**
冬天水杉林荫道

图 4-07　**Treegrove plaza in the fall**
秋季的林荫广场

间的人性化设计中，需要提供不同的游览路线、主题活动区域、共享开放空间和私密社交区域，以适应不同人群的社交需求。第一，提供丰富的游览路线。蜿蜒的路径可以让步行者欣赏不同的植物，视觉变化丰富的场景也为社交提供可分享的话题。步行道路的宽度应可容纳一人或多人同行，同时不会干扰路边休息的人群。第二，营造户外餐饮空间。户外就餐和茶歇是家庭或朋友聚会的重要活动。在公园和广场的社交空间中提供自主野餐的场地可以拉近家庭和团体的距离，提升社交的活跃度。在场地设计中可结合环境特征提供适宜轻餐饮或烧烤的户外家具，或者提供野餐活动的公共草地。第三，提供可产生对话和交流的场所。除了家庭或者约会人群相约去公园或者广场，还有部分人群是独自来到公共空间休闲的，这部分人群

包括刚下班的年轻人、来晒太阳的老人、刚买完菜的家庭主妇、附近出来喝咖啡的白领等。这些人群往往没有特别的社交需求，没有目的的闲逛是他们放松精神的主要活动。对于这部分人群，设计师的人性化设计也应提供相应的理解和关怀。对于不想被打扰的闲逛者，可以将部分路线以多层灌木分割，形成"小径通幽"的景观，或者利用地形变化形成相互能看见，但又互不打扰的区域。另外，也可为闲逛者提供休闲长椅、网球场、篮球场、遛狗区域等特定的公共设施，方便无目的的闲逛者自主选择想参与的活动，并以开放的形式展现给大家。

世界上受欢迎的城市公园和广场都具有较强的包容性，其特征在于对不同人群需求的满足。[79] 尽管设计师不可能同时满足所有人，但是可以通过空间的营造来创造不同的基础场

For wanderers desiring uninterrupted solitude, sections of the pathway can be demarcated by multiple layers of shrubbery, creating traversable routes that traverse the area. Alternatively, topographical variations can generate areas where visual contact is maintained without disruption. Benches may also be provided to afford wanderers respite, while specific public amenities such as tennis courts, basketball courts, and dog-walking zones offer aimless wanderers the freedom to select their preferred activities.

Popular urban parks and squares worldwide epitomize inclusivity, striving to fulfill the diverse social requisites of their denizens.[79] While it is impossible for designers to cater to every individual's needs simultaneously, they can establish a foundational framework through spatial configurations, allowing users to freely interpret and engage with the various scenes.

Design requirements for infrastructure

Considering the public nature of urban parks and squares, it becomes necessary to provide fundamental services that cater to users' needs. Signage, ramps and staircases,

parking facilities, illumination, waste receptacles, drinking fountains, public restrooms, and outdoor furniture (seating arrangements and children's play equipment, among others) constitute indispensable amenities in urban parks and squares.

In the first stages of planning, the consideration of single-use facilities' placement is necessary. The proximity of drinking fountains, public toilets, and seating arrangements must be taken into account. These three amenities are the most frequently utilized facilities within public spaces and are of utmost physical necessity. In addition to ensuring visual accessibility, the designer must address the demand for shade and wind protection by tailoring the design to the specific requirements of each site. For instance, the strategic positioning of trees and plants can serve as natural shade, while the construction of pavilions and small rest stations can offer additional services. The infrastructure should be designed in a manner that shields users from direct sunlight while utilizing these services and protects them from wind and rain while waiting to use the restroom. In addition, it is essential for these facilities to adhere

景，供使用者自由演绎。

（3）对基础设施的设计需求。城市公园和广场的公共空间需要提供基础服务设施以满足使用者的需求。其中标识牌、坡道和楼梯、停车场、灯具、垃圾桶、饮水机、公共厕所、户外家具（座椅和儿童游乐设施）等基础设施是公园和广场的必需品。首先考虑单体使用设施的位置规划。饮水机、公共厕所和座椅是公共空间最为常用并且符合生理需求的基础设施，应设在方便到达并且容易发现的位置。除了在视觉上容易发现外，遮阴挡风的功能则需要设计师针对不同场地的情况进行设计，让使用者在使用服务设施的过程中不会受到阳光的暴晒，在下雨时不会因为等待使用洗手间而受到风雨的侵扰。例如，可以使用树阵起到自然的遮蔽作用，或者修建凉亭或小型休息驿站提供服务。另外，设施需要考虑无障碍设计。以饮水机为例，饮水机不仅需要符合正常人体取水的高度，也需要照顾到弱势群体如儿童、坐轮椅的老人等的取水高度，减少取水不便的情况。座椅的设计需要考虑材料是否耐用以及是否符合人体工学。最后，基础设施设计还需要考虑可持续环境的建设。饮水机的供水和排水设施需要满足节水节能的可持续理念。例如，饮水机的出水系统需要考虑一次性喝水的出水量，没有节制的出水会造成浪费。溢出的水量也需要考虑是否可便捷地到达废水收集处，不会造成地面漏水过多的情况。同样，以可持续理念去思考灯具、发光标识牌、可分类垃圾桶的使用，都可以使基础设施设计更人性化、更合理。

accessibility. Taking drinking fountains as an example, their height should be suitable for children and individuals using wheelchairs. The durability of the materials and the ergonomic design of the seating arrangements must also be taken into consideration. Lastly, it is crucial to evaluate whether the infrastructure can be fortified to promote sustainable development. The water supply and drainage systems of the drinking fountains should align with the principles of water and energy conservation. The drainage system should account for the volume of water discharged at any given time, as uncontrolled discharge can result in excessive water wastage. Additionally, the overflow from the distributor should consider the accessibility of the wastewater collection point without causing excessive ground leakage. Similarly, the utilization of lamps, illuminated signs, and recyclable waste bins can be designed in a sustainable manner to create a more human-centric and rational infrastructure.

4.2.2 Human-centered city blocks
A high-quality neighborhood environment epitomizes the essence of human-centric design, serving as a pivotal space for effectively harnessing urban infrastructure and enhancing the quality of life.

Demand for natural environment design
The natural environment in a city block comprises the presence of street trees, urban greenways (including median strips, pocket garden greening, waterfront greening, etc), the courtyard greening, and planting ponds adjacent to the shops along the streets. Street trees play a pivotal role in shaping the overall ambiance of the community. The changing colors and forms of the trees throughout the seasons provide a dynamic landscape. As depicted in Figure 4-05, the emergence of new buds in spring embellishes the streetscape, infusing vitality into the surrounding areas. The resplendent street trees, adorned with early cherry blossoms, offer people an opportunity for celebration. The autumnal foliage of ginkgo trees imparts a warm and romantic atmosphere during colder weather. Even in winter, the fallen leaves and barren branches create a uniform landscape (Fig 4-06, 4-07) . These trees visually depict the passage of seasons and symbolize the natural chronology. On another note, street greening serves as a natural

2. 人性化街区

街区是城市的骨骼，也是展现城市基础设施使用效率以及生活品质的重要空间。高质量的街区环境是人性化最为突出的体现。

（1）对自然环境的设计需求。街区的自然环境营造，主要体现为行道树、交通绿道、街区庭院绿化、沿街商铺的种植池等。其中，行道树是街区整体气氛的重要组成部分。行道树的种类和四季落叶形成的树形姿态，为城市生活提供着四季变化的景观。春天的新叶点缀街道，为街区带来蓬勃的朝气。早樱的盛放美景，为人们带来节日的气息。秋天银杏树的落叶为寒冷的天气增添温暖浪漫的气息。即使是冬天的枯枝与建筑所形成的投影关系，也可以形成和谐的构筑画面。行道树为街区提供了季节变化的视觉效果，也是自然时间轴线的象征。

另外，街道绿化为城市道路的地表径流提供过滤作用。街道绿化以雨水花园的设计方式，使绿化区域形成自然下凹的地形关系，并通过种植具有净化功能的低维护植物品种，长期为城市道路提供雨水净化的作用。

（2）对社交环境的设计需求。街道是城市生活不可忽略的空间场所，也是市民生活的主要场景之一。人性化的街区环境，并非只有舒适的步行体验和干净的街道，同时必须具备便捷的功能服务和活力。富有生活气息的街道环境，往往伴随着丰富的商业、娱乐、休闲活动。市民可以通过走出家门，与邻居拉近关系，享受最近的公共休闲活动。对于连排形建筑来说，街道两边为商铺和车行道路，街区的社交场所不仅是街道区域，更以商铺为核心延伸到室内。日常杂货铺、面包店、咖啡厅、服装店

filtration system for urban road surface runoff. Adopting the form of rain gardens, street greening establishes a natural topographical relationship with green spaces. By planting low-maintenance plant species with purification characteristics, these rain gardens provide long-term purification of rainwater for urban roads.

Demand for social environment design

The street, an integral component of urban life, assumes a paramount role as the primary backdrop for citizens' daily existence. A neighborhood that revolves around the human experience comprises not only the pleasurable act of strolling amidst a pristine landscape but also the provision of convenient amenities, services, and a vibrant ambiance. A bustling street environment, teeming with vitality, is invariably accompanied by a profusion of commercial enterprises, recreational pursuits, and leisurely engagements. It affords denizens the opportunity to venture beyond the confines of their abodes, acquaint themselves with their neighbors, and revel in the proximity of public leisure spaces. Confronting this assemblage of buildings, flanking the thoroughfare, lie an array of shops

and alleyways. The social expanse of the neighborhood extends not solely to the street precinct but also extends indoor, with the shops serving as the epicenter. Daily provisions emporia, bakeries, coffeehouses, apparel boutiques, and jewelry establishments are interspersed among their ranks. While the residents are bereft of park-like seating arrangements, various shops have created social spaces. In the case of enclosed blocks, the greening of the block comprises the courtyard. This communal green space, while accessible to the public, is demarcated from the external thoroughfare, thereby producing a more secluded and tranquil leisurely green space. The neighborhood courtyard proffers the nearest public space wherein residents may enjoy. It also serves as the principal social arena for children and the elderly to frolic and unwind. To this end, it is necessary to foster an environment that is both secure and intimate, thereby ensuring that the activities of the children remain undisturbed. Conversely, the strategic placement of leisure benches or convivial tables provides the possibility of organizing diverse community-oriented events.

首饰店等都穿插其中，通过商铺的业态变化可以形成不同的社交场所。而对于围合型的街区，街区绿化包括了街区庭院。街区庭院为公共绿地，但与外部车行道路相隔，形成较为私密和安静的休闲空间，是离居民生活最近的公共空间。街区庭院也是儿童及老人休闲娱乐的主要场地之一。街区庭院的设计需要营造安全具有私密性的空间气氛，保护儿童的活动不被打扰。另外，应放置休闲长椅或者可以聚会的餐桌，为不同社区活动的举办提供可能。

（3）对基础设施的设计需求。街区的基础服务设施有自行车停放架、路灯、栏杆、垃圾桶、道路指示牌等。根据街区的宽度变化，基础服务设施的种类也各有不同。路灯、栏杆、垃圾桶等设施在所有宽度的街区都是必需品。对于商铺较多，步行宽度小于 1.5 米，以通行

为主的街区，自行车停放架无法匹配，在步行宽度 2 米以上的街区则可以设置。另外，部分尺度较大的、步行宽度达到 3 米及以上的街道，可以定义为商业步行街。[80] 这类街道对于基础服务设施的丰富性要求更高，如休息座椅、街道导览地图，以及可以短暂停留的饮水或充电区域等。街区的人性化服务设施需要根据街区环境和周边业态灵活配置，以适应不同人群的使用需求。

3. 人性化户外活动空间

根据建筑功能的不同，所对应的公共空间所呈现的人性化需求也各有不同。以校园户外活动空间、老年户外活动空间、儿童户外活动空间、医院户外活动空间四个主题进行人性化设计策略的分析。

（1）校园户外活动空间。古希腊语

Design requirements for infrastructure

The infrastructure of the community comprises bicycle parking racks, street lamps, railings, waste receptacles, and road signage. The nature of these infrastructural elements varies contingent upon the width of the block. Irrespective of the block's dimensions, street lamps, railings, and waste bins constitute indispensable amenities. In blocks characterized by a profusion of shops, path widths seldom exceed 1.5 meters, rendering the accommodation of bicycle racks unfeasible. However, pedestrianized street spaces exceeding 2 meters in width can readily accommodate such provisions. Conversely, expansive thoroughfares boasting pedestrian widths of 3 meters or more may be designated as pedestrian-only commercial streets.[80] These thoroughfares necessitate a broader spectrum of fundamental services, such as seating areas for repose, potable water stations, or transient charging stations. The street facilities must exhibit adaptability and flexibility, seamlessly harmonizing with the environment and the surrounding commercial enterprises, so as to cater to the diverse needs of the populace.

4.2.3 Human-centered outdoor activity spaces

The requirements of public spaces, corresponding to the buildings, are contingent upon the purpose of the structure in question. Four types—campus outdoor activity spaces, elderly outdoor activity spaces, children outdoor activity spaces, and hospital outdoor activity spaces—have been employed to analyze human-centered design strategies.

Campus outdoor activity space

The term college originates from the ancient Greek word *akademeia*, denoting a "sacred grove"[81] where religious and academic exchanges took place in the gardens of ancient Greece. In contemporary usage, this English term also comprises the concept of "academy." The ethos of academy, which emphasizes exploration and communication, resonates with the ancient Greek notion of a "sacred grove"—a communal space where ideas could freely intermingle and inspire. Therefore, the outdoor areas adjacent to the campus strive to optimize opportunities for interaction between students and faculty members. Campus planning can be likened to the planning of a miniature city, with expansive sports fields, commemorative plazas, and "akademeia" 的意思为"圣林"[81]，是古希腊进行宗教传播和知识交流的场所。在现代，"academy" 作为英语的"学术"一词，也代表着对大学的称呼。大学精神以自由探索交流为核心，呼应了古希腊文化对于"圣林"的释义——一个可以自由交流、激发想法的公共场所。因此，校园的户外活动空间设计应追求最大化地促进师生之间的交流。校园规划犹如小型城市规划，大型运动体育场、纪念性广场以及连接教学楼的街道空间是校园户外空间的主要构成部分，但在设计时往往对建筑前厅的公共空间有所忽视，而这种过渡性的前廊空间建构了学生上学期间的等待、聚会、聊天、讨论等活动时间（图 4-08）。

美国加利福尼亚大学伯克利分校曾经进行校园调查研究，出乎意料地发现让学生有归属感的场所是有着对应的桌椅或者可以进行固定讨论的场所。虽然学生经常穿梭于不同的教学楼进行学习，各种活动也会在不同学院里举行，但令学生产生强烈归属感的是与家相似的空间。这种与家相似的空间，可以理解为可自由席坐、聊天、工作的让人觉得轻松和愉快的场所。因此校园设计中的"过渡性空间"[82] 不应被简单地处理为大厅或者广场。基于人性化的设计考量，这些过渡性公共空间应为学生或者教职工提供讨论、交谈的场所。例如，提供既能服务于健全学生群体，又能服务于轮椅使用群体的并排或者圆桌的餐椅。在靠近教室入口的公共空间可提供饮水机、垃圾桶，或者可充电的座椅供短暂的休息讨论。在建筑入口或者学院门口可提供便利店、咖啡厅或者自动贩卖机，为学生提供短暂的能量补给。同时，景

thoroughfares linking academic buildings constituting the primary constituents of these outdoor spaces. However, the public spaces in building vestibules are often overlooked. However, it is precisely in these transitional spaces that students congregate, arrange meetings, engage in conversations, and deliberate throughout their academic day (Fig 4-08). A campus survey conducted by the University of Berkeley yielded an unexpected finding: the locations where students experience a sense of belonging are those where they can convene and converse. While students frequently traverse various buildings for their studies, diverse social events occur in different colleges. Nevertheless, it is the spaces proximate to their residences that provide a profound sense of belonging among students. These homely spaces can be conceived as areas where individuals can congregate, converse, work, experience joy, and unwind. The transitional spaces[82] in campus design should not be regarded merely as corridors or squares. In the context of human-centered design, these transitional public spaces should offer venues for students and staff to engage in discussions, dialogues, and knowledge-sharing. For instance, seating arrangements can accommodate both able-bodied students and those

utilizing wheelchairs, with dining chairs arranged side by side or round tables provided. Public spaces near classroom entrances can feature drinking fountains, waste receptacles, or seating areas for brief respites and discussions. Convenience stores, cafes, or vending machines can be situated at the entrances of buildings or community colleges. Outdoor benches or seating arrangements for communal lunches can also be positioned in areas featuring picturesque vistas. These considerate and user-friendly amenities can enhance the well-being of students on campus.

Elderly outdoor activity spaces
The human-centered design of outdoor spaces for the elderly necessitates attention not only to physical safety but also to psychological needs. With regard to physical requirements:

• Transitional spaces from buildings to the outdoors should possess a semi-enclosed configuration to shield against inclement weather. If the building vestibule is sufficiently spacious, seating arrangements can be provided to facilitate rest and contemplation.

• Entrances and corridors necessitate handrails to assist

观视线较好的区域可提供户外休息的长椅或者用餐聚会的座椅。这些贴心的人性化服务设施可提升学生校园生活的幸福感。

（2）老年户外活动空间。对老年户外活动空间的人性化设计不仅需要关注人身安全，更需要体现环境对于心理需求的回应。

针对物理需求的设计：

建筑到户外的过渡地带，需要设计可以挡风防雨的半围合空间。如果建筑前厅足够大，可以提供休息观景的座椅。

在建筑出入口和走廊区域需要安装扶手，为老人散步和停顿提供支持。特别是建筑出入口处需要设置无障碍通道，方便使用轮椅的老人出行。

户外活动空间的铺装需要使用防撞的软质地垫以及防滑材料，以免老人雨天走路摔倒受伤。

户外活动空间的步行道与住宅入口需要有清晰的指向性，避免老年人困惑和迷路。

户外活动空间面向开阔的场地，以免建筑或者高大树木的遮挡形成视觉死角，增加老人受伤而没能及时被发现的风险。

针对心理需求的设计：

提供可小型聚会的户外活动场所，设置餐椅和遮阳棚，以便老年人享受阳光和自然风景（图4-09）。

提供辅助式运动设施，如步行道、坡道和拉伸器械，鼓励老年人进行适当难度的户外运动。

增加可以观赏的园艺区域，种植居民可以参与维护的植物，丰富老人户外活动的内容。

老年人户外活动空间需要确保安全性、私

the elderly in walking and maintaining balance. Particularly, wheelchair accessibility is necessary at building entrances.
• Outdoor areas should be covered with soft, impact-absorbing mats and non-slip materials.
• A clearly demarcated pathway leading to the entrance of the residence is indispensable to prevent confusion.
• The outdoor space should face open areas, avoiding obstruction by buildings or towering trees, so as to prevent blind spots and minimize the likelihood of elderly individuals sustaining injuries without immediate assistance (Fig 4-09).

For the design of psychological needs:
• Outdoor spaces should incorporate areas for small gatherings, complete with outdoor dining chairs and awnings, enabling individuals to bask in the sunlight and relish the natural scenery.
• Supplementary facilities should be provided to facilitate and encourage outdoor physical activities for individuals with limited mobility. Examples include sidewalks and ramps.
• Enhancing ornamental gardening activities and

cultivating greenery serve to enrich the outdoor experience.
• The outdoor activity spaces for the elderly necessitate the assurance of safety and privacy, thereby instilling in them a profound sense of security conducive to their engagement in outdoor pursuits and fostering their active involvement in social activities, thereby enhancing their overall comfort and enhancing their quality of life.

Children outdoor activity spaces

The design of outdoor spaces for children's activities is predicated upon the consideration of three fundamental aspects: the scale of children's requirements, safety, and the educational environment. From the vantage point of children, all recreational and urban facilities should be tailored to the height and visual acuity of children aged three to five. The placement of children's facilities amidst towering arboreal specimens or dense bushes results in the potential danger of children succumbing to falls or crouching whilst engaged in locomotion or play, thereby impeding the line of sight of their supervisory adults. From a safety standpoint, outdoor spaces necessitate an orientation towards unobstructed thoroughfares or unimpeded structures,

图 4-08　**Outdoor steps at Columbia University, USA**
美国哥伦比亚大学户外台阶场景

图 4-09　**Outdoor party for senior citizens**
老年户外聚会活动

密性和流通性，给予老年人足够的安全感进行户外活动，并鼓励老年人社交，从而增强精神上的慰藉。

（3）儿童户外活动空间。儿童户外活动空间的设计有着较强的特殊性，主要体现在儿童尺度需求、安全性、幼托环境的教育性三个方面。在儿童的尺度需求上，所有娱乐设施和生活设施都需要按照三到五岁儿童的身高以及视线进行设计布局。如果活动设施设置在较大的乔木附近或者茂盛的树丛中，则可能产生视觉盲区，小孩在奔跑或玩耍中摔倒或蹲下时成人的视角无法及时观察到。在安全性上，户外活动的空间需要面向开阔的街道或者无构筑物遮挡的区域，并且需要有栏杆围合。街道步行的路人以及园区的老师可以对活动场地进行双向监督，避免儿童在游玩过程中出现意外。特

别是儿童游玩设施附近应提供长椅，使照看儿童的家长在休息的同时可以关注儿童动向。此外，还需要给幼龄儿童提供可探索的环境，运用环境进行智力开发。包括户外设施使用可以刺激儿童感官的材质和设计，如利用树叶、花卉、铺地材质的变化来丰富游玩的体验。在植物配置中，可增加带有花果或者香味的植物品种，便于儿童观察自然变化，发现自然规律（图4-10）。

（4）医院户外活动空间。医院户外活动区域的使用者大部分是住院疗愈的病人，空间功能是使病人在康复的过程中能更好地进行户外锻炼。由于医院紧促的建筑面积，大部分户外活动空间都以绿地和简单的座椅进行布局。但医院的户外活动场所需要关注流线的安全性以及对医疗的辅助作用。医院的户外活动空间

encircled by protective railings. Pedestrians traversing the thoroughfare and teachers in the park can effectively monitor the space from both directions, thereby averting the occurrence of injuries to children that might otherwise go unnoticed. Notably, benches may be strategically positioned in proximity to children's play equipment, thereby affording parents an environment conducive to relaxation whilst simultaneously enabling them to maintain a vigilant watch over their progeny's activities. In addition, children's outdoor spaces should provide an environment that facilitates exploration and developmental growth. This involves the incorporation of sensory-stimulating materials, such as foliage, blossoms, and ground coverings, thereby enriching the experiential dimension of play. In terms of plant configuration, the inclusion of species bearing flowers, fruits, or fragrances serves to provide activities that enable children to observe the vicissitudes of nature and discover its patterns (Fig 4-10).

Hospitals' outdoor activities areas

The majority of outdoor areas in hospital premises are designated for the use of in-patients, thereby affording

them the opportunity to engage in outdoor exercise during the course of their convalescence. In light of the scarcity of hospital spaces, the outdoor areas in hospital precincts predominantly comprise verdant expanses and unadorned seating arrangements. However, outdoor spaces intended for medical care must duly prioritize considerations of movement safety and their therapeutic impact on patients. To this end, outdoor spaces in hospital premises should offer patients handrails for ambulation or seating arrangements for brief interludes of interaction. The design of seating arrangements must duly account for the diverse abilities of individuals. For instance, adults utilizing wheelchairs and children reliant on wheelchairs necessitate outdoor seating arrangements of varying heights. Elderly individuals or children utilizing crutches must be able to access water fountains. Moreover, outdoor spaces in hospital premises must also cater to the movements of personnel, medical staff, and visitors. Human-centric design principles are not solely applicable to individuals with mobility impairments; rather, all individuals frequenting and working on hospital premises should derive solace from the outdoor spaces. Functional zoning facilitates

需要设置供病人在散步过程中停留的扶手，或者可以休息交流的座椅。座椅的形态需要考虑各种行动不便的人群，如使用轮椅的成人与使用轮椅的儿童需要的座椅高度不一样。另外，医院的户外空间需要照顾到医疗医护人员和探病家属的活动流线。人性化设计并不是只为患者提供使用便利，医院中的所有工作者和来访者都有享受户外空间的权利。在功能分区上，可以设置不同方向的动线，让医务人员能便捷地走到户外空间进行休息而与病人活动的空间有遮挡距离。探病的家属也能与患病的家人享受自然环境的美好而不受打扰（图4-11）。医院户外活动空间的设计方法有很多，人性化设计应以不同人群的视角进行观察，才能在环境的营造中照顾到患者和非患者的需求。

三、无障碍设计

20世纪初，由于人道主义的呼唤，建筑学展开了无障碍设计的讨论。无障碍设计秉承弱势群体、残疾人以及健全人都享受"平等"对待的设计原则。[83]在具体实施上，无障碍设计主要针对残疾人、老人、病人、儿童等，减少其在户外活动的障碍和不便。

无障碍设计在城市规划中主要体现为以下三个方面。第一，通过指示系统和基础设施辅助残疾人步行。例如，在道路步行系统中设置盲人步道、可语音按键的红绿灯系统、可触摸的盲人地图或者智能语音导览等，辅助残疾人自如在城市中通行，并受到相应的安全保护。第二，建立残疾人专用的卫生间、符合无障碍设计规范的坡道及栏杆、公用电话、具有语音引导的取款机等公共服务设施。第三，在娱乐

different pathways, thereby enabling medical personnel to conveniently access outdoor courtyards for respite away from patients. Families visiting patients may likewise bask in the splendor of the natural outdoor environment without undue disturbance (Fig 4-11). Numerous approaches exist for the design of outdoor activities on hospital premises. The adoption of a human-centric design ethos necessitates the consideration of diverse perspectives, thereby ensuring the continued fulfillment of the needs of patients and staff in the creation of the environment.

4.3 Accessible design

At the beginning of the twentieth century, the call for humanitarianism instigated a profound discourse concerning the architectural configuration of accessibility. The concept of accessible design is rooted in the fundamental principle of rendering spaces and activities reachable and utilizable by all individuals, including the disadvantaged, the disabled, and the able-bodied.[83] The essence of barrier-free design lies in circumventing the impediments that individuals with disabilities, the elderly, patients, and children may encounter in outdoor

environments. The urban planning encapsulates three primary facets of barrier-free design.

First and foremost, in urban planning, it is necessary to cater to the requirements of individuals with disabilities through the implementation of signage and infrastructure. Illustrative instances encompass unobstructed pathways for the visually impaired, a system of traffic lights equipped with voice-activated buttons, tactile maps tailored for the visually impaired, and audio guides facilitating the comfortable navigation and provision of appropriate protection and assistance to individuals with disabilities.

Secondly, the availability of public amenities such as restrooms, ADA-compliant ramps and railings, public telephones, and cash machines equipped with audible navigation is of paramount importance.

Thirdly, the provision of recreational and leisure spaces for individuals with disabilities is indispensable, thereby enabling their active participation in recreational pursuits. The barrier-free environment also takes into

图 4-10 **Children's Museum in Sonoma County**
索诺玛县儿童博物馆户外教育设施

图 4-11 **Royal Stoke University Hospital**
皇家斯托克大学医院

account individuals with children, individuals with limited physical mobility, and individuals afflicted with various chronic ailments. In addition to fundamental facilities, healthcare establishments adhering to barrier-free design standards can enhance the social engagement of disabled individuals, thereby bolstering their overall well-being and mental health. For instance, Tokyo Disneyland has devised an accessibility plan for the park, including a comprehensive guide to the park, theater show schedules, and themed promotions employing audio and disability signage. Tactile Braille maps and staff proficient in sign language are made available to facilitate communication (Fig 4-12). The park is also thoughtfully designed to incorporate priority restrooms for disabled individuals, fountains with appropriate heights for wheelchair users, and entrance and exit ramps equipped with handrails. Barrier-free design principles have been assimilated into the standards governing housing and urban planning across diverse countries. As a designer, it is incumbent upon oneself to adhere to these standards to ensure the quality of design, while simultaneously recognizing that there exists no singular optimal approach to constructing

a barrier-free design. Only through incessant observation and optimization can the perpetual advancement of superior barrier-free environments be achieved.

4.4 A human-centered environment

In addition to the planning and spatial design aspects, the pursuit of human-centered environmental design necessitates an exploration of individuals' desires for their surroundings. The creation of a human-centered environment comprises not only the aforementioned planning and spatial design elements but also the provision of high-quality services tailored to the users of said environment. Designers must possess a profound understanding of life and possess the capacity to observe and analyze issues, enabling them to continuously devise innovative solutions. How can we effectively implement a human-centered environment? The answer lies in tailoring solutions to the unique cultural and environmental characteristics of each city. However, by incorporating community participation into the design process, all parties involved can gain a deeper understanding of one another's perspectives (Fig 4-13).

和休闲空间提供残疾人也可参与游玩的设施。无障碍设计服务对象也包括带着孩子游玩、身体机能低下，以及具有各方面慢性疾病的人群。除了基本生活设施，还可以建立符合无障碍设计标准的康健设施，增加残疾人群体的社交活动机会，从而带来精神上的提升。例如，东京迪士尼乐园在游园的无障碍规划中，通过语音和残疾人标识展示游园指南、剧场演出时间、主题宣传等。针对视觉障碍的人群，提供可触摸的盲文地图并配置手语工作人员辅助交流（图4-12）。此外，还提供残障人士优先的卫生间、符合轮椅使用者高度的饮水机、有扶手的出入口坡道等，采用全方位的辅助设计。无障碍设计规范已被不同国家纳入住宅和城市规划设计标准。设计师需要依循设计规范，保证设计质量，不断观察和优化，更好地进行无障碍设计。

四、人性化的服务环境

人性化的环境设计，除了规划和空间层面的建设，也需要从人的管理自治的视角，探索人性化服务对于环境质量提升的意义。设计师需要具有较强的生活感受能力以及观察问题、分析问题的能力，从而可以不断提出书本中没有的解决方法。人性化服务环境的提升应该如何讨论和执行呢？尽管城市文化和背景的不同会指向不同的解决方法，但是我们都可以在设计的流程中加入社区居民的参与（图4-13），让双方能从各自的角度了解对方。只有设计师脱离"我认为"的视角，真正从居民角度思考解决方案，人性化服务的意义才会在社区里被真正践行，并不断得到维护和巩固。

Only when designers are able to consider solutions from the standpoint of actual residents can they genuinely provide superior human services to the community.

The design workflow can be divided into four primary stages: site investigation, design problem analysis, scheme design, and construction supervision.[84] During the site investigation phase, the designer administers a comprehensive questionnaire and evaluates the site based on its dimensions, the requirements of the surrounding residents, and the environmental quality. Nevertheless, due to subjective preferences, many questionnaires fail to accurately target the genuine needs of the residents. Therefore, numerous proposals may appear impressive on paper but appear unused by the residents once constructed, leading to abandonment or redevelopment. This predicament typically arises from a lack of effective communication during the design process. The residents are unable to understand the correlation between the design and the final environment from the designer's perspective, while the designer struggles to ascertain the residents' most pressing environmental needs.

Two research methods have proven particularly effective in bridging the gap between residents and participatory design research. The first method involves conducting public expert reviews and work-in-progress presentations. During the expert review session, landscape architects, architects, planning departments, and developers present a professional analysis of the site, a prototype of the design challenges, and a detailed discussion of the renovation's design options. The public is provided with a comprehensive and professional analysis of the drawings and the design's implications, thereby offering the city's residents a reference point and an understanding of the pre-design and future development trends[85] (Fig 4-14). This establishes a foundation for mutual communication. While this research method is commonly employed by universities and professional design firms, it predominantly adopts a macro perspective from the designer's vantage point. Urban residents are deprived of the opportunity to express their desires to designers and relevant government agencies. The second research method involves organizing seminars and scenario planning sessions, facilitated by government planning funds or community committees.

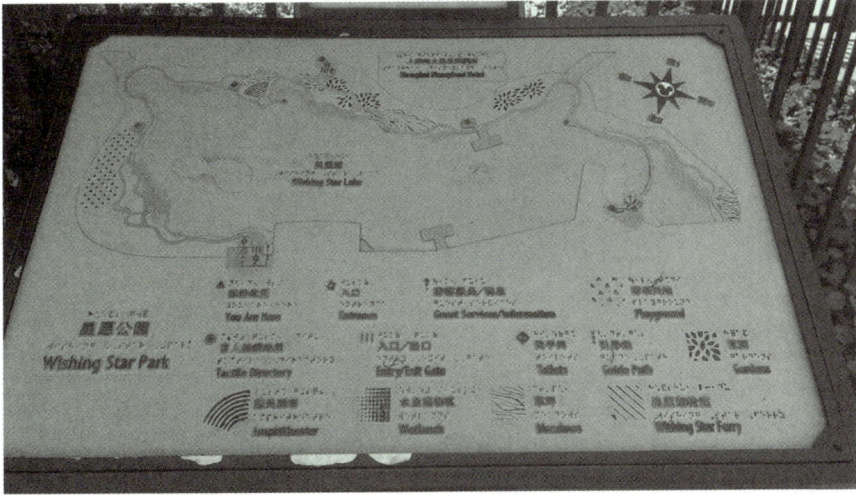

图 4-12 Touch-responsive 3D maps in Tokyo Disney
东京迪士尼的固定触觉地图

图 4-13 Resident meeting
居民沟通会

设计的工作流程可分为场地调研、设计问题分析、方案设计、施工监理四大部分。[84] 在场地调研中，设计师以场地为对象，对场地尺寸、周边居民需求、环境质量进行问卷调查和测评。但是大部分的问卷调查有主观偏好的影响，无法很好地定位居民真正的需求，从而导致很多方案的设计效果图看起来非常漂亮，但是实际建造完成后居民无法很好地使用，最后导致荒废或重新改造。这样的结果大多是因为在设计的过程中没有进行有效的沟通。居民无法从设计师的角度理解设计对于未来环境提升的意义，而设计师也未能理解居民最迫切需要解决的生活需求是什么。

在调研中，有两种提升居民参与度的研究方法是十分有效的。第一种是公开的专家评审和工作进度展示。在专家评审会上，景观设计师、建筑师、规划人员以及开发商就项目的改造设计进行专业的地块分析、设计挑战模拟以及方案细节讨论，向公众呈现专业完整的分析图纸并介绍设计意义，帮助居民获得参考，了解设计前期情况和未来发展动向，从而为彼此的沟通建立基础[85]（图4-14）。这是高校以及大部分专业设计公司常用的调研方式，不过这种方法还是从设计师的宏观角度看待问题，居民没有机会向设计师和相关政府机构提出他们的意愿。第二种方法则是通过政府规划基金会或者社区委员会开展工作坊探讨和情景规划，通过"自下而上"的方式，让居民有机会参与设计的全过程，从而辅助设计团队更为全面地解决问题，增强居民对新环境的联系感。

上海长宁区的乐颐生境花园[86]（图4-15）是设计师与居民共建的案例之一。小区内不到

This approach empowers residents to actively participate in the design process through a "bottom-up" approach, thereby assisting the design team in comprehensively addressing the problem and fostering a stronger sense of connection between the residents and the regenerated environment.

The Habitat Garden, located in the Changning District of Shanghai[86] (Fig 4-15), exemplifies joint efforts between designers and residents. This park, spanning an area of less than 100 square meters, has garnered recognition as one of the world's top 100 exquisitely crafted gardens. The garden's maintenance constitutes a long-term project undertaken by the residents. Active involvement from the residents was observed throughout the design process, including plant selection and cultivation. The residents derived a profound sense of fulfillment from their contributions to the garden's construction and the subsequent enhancement of their living environment, thereby fostering a greater dedication to upholding environmental quality. Similarly, the successful design of the High Line Park in the United States also reflects the significance of the design process. A dedicated community committee was established to address communal concerns. During the early stages of design, the committee actively sought input from neighboring residents, employing seminars to simulate the residents' perspective and deliberate potential design challenges. Following the park's completion, the committee devised conventional routes for plant observation, organized musical activities, and crafted stories that explained the park's design. By fostering collaboration between the community and designers, the paradigm of human-centered design can be broadened, thereby forging a closer bond between the community and the urban environment.

1000 平方米的绿地公园却被评为"世界上 100 个最美的生境花园"之一，与后期花园的维护和居民长期的悉心照顾脱离不了关系。从设计方案到植物的选择和种植，在设计的全过程中都有居民的参与和共建。居民对于花园的建设和居住环境的提升有较大的成就感，所以对于维持环境的质量也变得更为用心。另一个例子就是美国的高线公园。在高线公园的设计过程中以及建成后，都有专门的社区委员会进行维护。在设计初期，社区居委会征集了周边居民的意见并进行回访，以工作坊的方式，通过居民的视角进行设计模拟，一起探讨设计中可能遇到的挑战。在建成后，委员会定期开展公园植物观赏、公园音乐演奏、讲述设计故事等活动，不断为高线公园提供宣传的途径。公众与设计师共建的模式，让人性化设计的视野变得

更为广阔，也让公众与城市环境的关系变得更为密切。

小结

通过人性化设计的概念介绍，以及具体的设计实践转换，拓展关于人性化场所应用方式的认知。人性化设计不仅体现在物理环境的场所设计方面，还体现在人性化服务、管理运营、教育宣传等软性服务方面。

Summary

Through the introduction of human-centered design principles and the adaptation of specific design methodologies, a more profound understanding of the creation of human-centered spaces can be attained. It becomes evident that human-centered design extends beyond the physical aspects of environmental design, comprising the design of intangible services, management and operations, education, and policy.

图 4-14 Project presentation
方案展示

图 4-15 Shanghai *Leyi* habitat garden
上海乐颐生境花园

课后作业

After-class exercises

———

讨论题

Discussion

人性化设计对于城市公共空间的影响在哪里?

How does human-centered design affect urban public space?

设计实践

Design practice

在日常生活中,城市哪些空间呈现"非人性化场所"? 根据讨论题内容,结合观察的场地制作设计评价表。

In daily life, what types of spaces in the city can be considered "non-human-centered places"? Develop a design evaluation table based on theoretical issues and on-site observations.

推荐读物

Reference reading

1.Gehl, J. *Life Between Buildings: Using Public Space*. Island Press, 2011.

2.Kimmelman, M. *The Intimate City: Walking New York*. Penguin Press, 2022.

3.Manshel, A. M. *Learning from Bryant Park: Revitalizing Cities, Towns, and Public Spaces*. Rutgers University Press, 2020.

4.Sim, D. *Soft City: Building Density for Everyday Life*. Island Press, 2019.

5.Speck, J. *Walkable City: How Downtown Can Save America, One Step at a Time*. macmilan, 2013.

6.Steg, L. and De Groot, I. M. *Environmental Psychology: An Introduction*. John Wiley & Sons, 2018.

7.Whyte, W. H. *The Social Life of Small Urban Spaces*. Project for Public Spaces, 2001.

第五章

景观的空间尺度

Chapter 5
Landscape spatial scale

学习要点
Study Points

理解尺度既是一种物理条件，也是一种文化条件，它的变化取决于环境、时间和预期活动。

Understand that perception of scale is both a physical and a cultural condition that changes depending on context, duration, and intended activity.

学习目的
Teaching Objectives

了解尺度的客观和主观模式如何对景观设计产生影响，以及如何让使用者感受它。

To understand how objective and subjective modes of perceiving scale contribute to a landscape and how it is received by an inhabitant.

本章通过尺度、感知和人的体验探讨景观设计的基本问题，重点介绍一系列案例研究，以展示尺度如何从身体中产生以及人手中可以握住什么。本章还将探讨不同的景观类型，以激发学生在不同环境条件下的批判性思维和设计可能。

Both landscape and landscape architecture are contingent upon activities, physical character, climate, and context. This chapter explores fundamental questions about landscape architecture through the lens of scale, perception and people. The chapter focuses on a range of case studies to demonstrate how scale emerges from the body and what can be held in one's hand. The chapter explores different landscape typologies to instigate critical thinking and design opportunities in each condition.

美国作家、哲学家拉尔夫·沃尔多·爱默生在他的作品《自然》中写道："我们指的是多种自然物体所形成的印象的完整性……我今天早上看到的迷人风景，无疑是由大约二十或三十个物体组成的农场。米勒拥有这片土地，洛克拥有那片土地，曼宁则拥有远处的林地。但他们都不拥有这片景观。地平线上有一种属性，只有眼睛能整合所有部分的人，即诗人，才拥有这种属性。"[87]

爱默生的名言包含了景观的许多元素——什么是可见的？什么是连贯的？什么是被理解为在视觉上由边缘、范围和实践界定的——而不是由财产线界定的？"景观"这个专业名词可以指特定的场所、设计或想法的集合，如政治或意识形态景观。景观设计是一个领域和一种实践，设计我们居住和体验的开放空间，包括从小花园到从太空可见的基础设施。这些景观可能保留文化和社会习惯，或者引入新的文化和社会习惯。

用一句话来概括，景观和景观建筑的设计都取决于活动类型、自然特征、气候和环境的综合考量。

一、尺度

我们对周围世界的第一次测量使用的是我们的手和身体。这种通过身体的测量是我们对比例最初的理解。这个物体是比我大还是小？通过了解能拿什么和不能拿什么，我们学会了理解物理范围。随着成长，我们通过运动、听觉和视觉来理解尺度。我们对风景的感知是由距离和远近的概念构成的：脚下的、可听的、可见的。我们对尺度的感知也是有条件的。

The American writer and philosopher Ralph Waldo Emerson, in his essay, *Nature*, writes, "We mean the integrity of impression made by manifold natural objects… The charming landscape which I saw this morning, is indubitably made up of some twenty or thirty farms. Miller owns this field, Locke that, and Manning, the woodland beyond. But none of them owns the landscape. There is a property in the horizon which no man has but he whose eye can integrate all the parts, that is, the poet."[87]

Emerson's quote holds so many elements of a landscape—what is visible, what coheres, what is understood as being delimited visually by edges, extents, and practices—rather than by property lines. "Landscape" is a term that can refer to a specific locale, a design, or a collection of ideas, such as a political or ideological landscape. Landscape architecture is a field and a practice that designs open spaces that we inhabit and traverse ranging from small gardens to infrastructure visible from space. These landscapes may hold cultural and social habits or introduce new ones. Both landscape and landscape architecture are contingent upon activities, physical character, climate, and context.

5.1 Scale

Our first measure of the world around us is through our hands and our bodies. This physical measure is our first understanding of scale. Is this object larger or smaller than I am? From comprehending what we can hold and what we cannot, we learn to understand physical extent. As we grow, we understand scale through movement and perceiving sound and sight. Our perception of the landscape is layered with notions of distance and nearness: what is underfoot, audible, and visible. Our perception of scale is also contingent.
Frequently, a landscape architect creates a model to understand a site. This tactile, portable model becomes a way of organizing the site—understanding and forming its edges, aiding in creating its coherence. This model may begin as a physical double of the site, emphasizing its contours and its limits and quieting the systems running through a parcel.

Selecting a scale for a physical model is an act of design—a means to create a physical and visual proxy for the actual site. The scale model correlates to the measure of a

通常情况下，景观设计师通过创造模型来理解场地。这种可触摸、可携带的模型成为组织场地的一种方式——理解并形成其边缘，帮助创造其连贯性。这个模型可以作为场地的一个实体替身——强调它的轮廓和它的界限，并使整个地块的系统确定下来。

为物理模型选择一个比例是一种设计行为——一种为实际场地创建物理和视觉替身的方法。这个等比模型与景观的比例相关——但也可以成为景观的概念性替身——设计如何固定、如何流动、如何包容。模型是一个真实场所的替代品，以我们可以快速理解的比例来打造该场所。将基地缩小到可以握在手中，有助于景观设计师看到新的关系，而材料、植被所覆盖的场地上的细节可能会掩盖这些关系。模型减少了天气和时间造成的差异影响，可以成为创造新秩序的有效工具。等比例模型也可以重点突出需要强调的部分，或者弱化不重要的内容。

等比模型是一种与材料相融合的思想表现，它将物理条件与它们在真实场所中的相互关系联系起来。一个等比模型创造了地貌的形象，同时排除了贯穿该地貌、在其之下和之上的系统。虽然等比模型强调了秩序感和形象感，但它会掩盖一系列看不见却在发挥作用的政治和历史条件。等比模型代表了真实环境，并将大地和水体作为固定不变的元素。比例模型在一定程度上是有用的，但至关重要的是要时刻牢记，模型是虚构的，模型中的问题不代表真实面对的所有设计问题。

二、感知

美国伊姆斯夫妇创作了一个著名的等比模

landscape, but it can also be a conceptual double for it—how fixed, how fluid, how inclusive. Models are stand-ins for a real place that describe the site at a scale that we can comprehend quickly. Shrinking the territory to something that can be held in one's hands helps a landscape architect see relationships that may be obscured by the details on a material, vegetated site. The model diminishes the differences caused by weather and time of day and can be a productive instrument to help create a new order. The scale model can also exclude important, less visible conditions.

A scale model is a representation of ideas melded with material that holds physical conditions in relation to one another as they may in a real place. A scale model creates a figure of the landform while denying the systems that are running through, under, and over a site. While scale models privilege a sense of order and figure, they can elide a set of unseen political and historical conditions that set into motion potential actions. A scale model stands in for the actual and holds earth as a fixed element that does not move, bodies of water that do not change. A scale model is useful—up to a point—but it is crucial to actively remember that a model is fictional and must exclude more than it includes.

5.2 Perception

A famous scale model created by American couple Charles and Ray Eames is large enough to poke one's head in to see the space as one would at eye level. The Eames continued to play with notions of scale and made them visible to all through their *Powers of Ten* films which they wrote and directed. The two short documentary films contemplate scale and proximity starting with a couple on a picnic blanket in a park and zoom 10 times farther out every 10 seconds until we arrived at the then edge of edge of the universe, 100 million light years away from the starting picnic blanket. We then zoom back in via powers of 10 until we end inside a proton of a carbon atom within a DNA molecule in a white blood cell in one of the picnickers.

The films altered perception at the time because wild variations between zooming in and out were novel, but now we have computer models that allow us to change our sense of scale effortlessly—one can always further zoom

型，大到足以让人把头探进去，看到与视线持平的空间。伊姆斯夫妇持续玩弄尺度的概念，并通过他们自编自导的电影《十的次方》让大家看到了这些概念。这部影片以规模和距离为主题，从一对夫妇在公园里的野餐毯上开始，视野每隔 10 秒放大 10 倍，直到我们到达当时已知的宇宙边缘——离开始的野餐毯有 1 亿光年。然后再通过 10 次方聚焦，直到在其中一个野餐者的白细胞 DNA 分子中的碳原子质子内部结束。

这部电影改变了当时人们的感知，因为放大和缩小之间的剧烈变化是新奇的。但现在我们有了计算机模型，可以毫不费力地改变我们的比例感——人们可以进一步放大或缩小这些场景。当一个人在脑海中有一千米的风景时，当细节可以同时被解决时，比例和感知的概念

是如何改变的？在电脑上设计一个网站如何改变我们的感知和行动路线？

这种缩放的概念已经成为 Rhino 和 Revit 等软件的第二性，这些软件使我们可以在无法完全了解和计算物理比例的情况下进行设计。该软件隐藏这些信息，使一切保留下来的信息只与屏幕大小相关，与景观大小几乎无关。实体模型寻求有助于解决实际问题的比例。这些模型使具有可操作性的距离、接近度和感知的概念更加直观。我们仍需要在数字领域找到方法来帮助我们更直观地理解比例，也许虚拟现实、混合现实和增强现实技术可以成为这个领域的福音。

美国景观设计师丹·凯利设计并建造了福特基金会花园，他的设计使用刨花板来制作模型。这个模型可以在纽约进行调整，它代表一

in or out. How does the notion of scale and perception change when details can be resolved at the same time one holds a landscape of a kilometer in one's mind? How does designing a site on a computer change our perception and our course of action?

This notion of zooming has become second nature with software programs like Rhino and Revit that make it possible to design without fully understanding and reckoning with physical scale. The software hides this information by making everything relative to screen size, which has very little relation to the distances of the landscape. Physical models had intention related to scale that each scale had a particular problem-solving quality. These models helped make more intuitive the notions of workability distances, proximity, and perception. We will need to find ways in the digital realm to help us understand scale more intuitively. Perhaps VR, MR, and AR technology can be a boon in this realm.

American landscape architect Dan Kiley created the Ford Foundation Garden using a chipboard model that could

be reworked in New York in a space that measured 80 feet by 80 feet in internal space that rose 120 feet high.[88] This atrium garden provided a refuge for workers and members of the public from the concrete, stone, and glass of the building and surrounding city. This garden worked concurrently on two scales—one at the scale of an indoor civic room in the city and one at the scale of the body with small pockets to escape being so visible and present in public.

Ana Mendieta, a Cuban-American sculptor, wrote, "Through my earth/body sculptures I become one with the earth… I become an extension of nature and nature becomes an extension of my body."[89] Her *Siluetas* series inscribes her own body-scaled form through subtraction in the sand. Sometimes it is set on fire, sometimes filled with pigment; other times, the artist lies wrapped in these spaces. These absences suggest burial or emergence. Taking a full-scale exploration makes the work connect to each of us.

Splash Pad Park by American landscape architect, Walter Hood, occupies 1 acre in Oakland, California. The

个面积约为594平方米，高约36米的内部空间。[88]这个中庭花园为工人和公众提供了慰藉，让人们有机会远离建筑和周围城区的混凝土、石头和玻璃。这个花园体现两个尺度——一方面是城市中室内空间的尺度；另一方面，它为身体的尺度提供了小型私密空间，使人能避开公众视线，获得片刻安宁。

古巴裔美国雕塑家安娜·门迭塔写道："通过我的《大地－身体》雕塑，我与大地融为一体。我成为自然的延伸，自然成为我身体的延伸。"[89]她于1973—1977年创作《轮廓》系列，在沙子上雕刻下自己的身体比例。有时这一比例雕塑被火点燃，有时用颜料填满；其他时候，艺术家的身体被包裹在这些空间里。这一系列的缺失表现都暗示着埋葬或显现。通过全尺度的探索和表现，使作品与我们每个人联系起来。

美国景观设计师沃尔特·胡德的戏水公园位于加利福尼亚州奥克兰，占地约4046平方米。该公园横跨高速公路下的土地，重新激活了人行横道，将当地社区与自身连接起来，并为一个历史悠久的剧院提供了更合适的通道。这个公园的适度规模为当地的农贸市场提供了空间，并鼓励人们居住在高速公路下方和附近的空间。它低矮蜿蜒的墙壁其实是精心设计的路线，供人们见面、交谈和玩耍。

在更大的规模上，由阿兰·普罗沃和吉尔·克莱芒设计的法国巴黎雪铁龙公园在城市中创造了一个广阔的草坪，并被一条斜向铺设的人行道分割开来（图5-01）。这片草坪的尺度如同一座城市客厅，与巴黎人想象和体验中的其他草坪相连。

在这片草坪的两边是一系列部分封闭的花

park spans land that is under a freeway and revitalizes pedestrian crossings to connect the local community to itself and to provide more fitting access to a historic theater. This park's modest scale introduced space for a local farmer's market and welcomes people to dwell in the space under and near the freeway. Its low winding walls suggest lightly choreographed routes for people to meet, converse, and play.

On a grander urban scale, Parc Andre Citroen designed by Allain Provost and Gilles Clement in Paris, France, creates an expansive lawn in the city bisected by a paved diagonal pedestrian path (Fig 5-01).

This lawn is at the scale of a city room which connects to other lawns in the imaginations and experiences of Parisians. Lining this lawn are a series of partially enclosed gardens, each dedicated to a color or sense. These rooms provide a refuge and an intermediate scale. These rooms are more introverted landscapes helping us focus on the stillness we hold in ourselves. The Silverleaf Room acts as a transitional mediator between rural and urban landscapes.

This landscape moves us from a city condition to a grove and back again.

Central Park is a city-scaled courtyard for New York City. It physically delimits streets with sunken east–west roads to create picturesque views and uses the surrounding edge of buildings to create a giant quadrangle among buildings from 57th to 110th Streets. The park organized its meadows separated by winding paved and unpaved paths. The largest uninterrupted spaces are given over to bodies of water. A formal path suggests movement and exchange with one another and with the trees lining the path. The park anticipated the development of buildings along its perimeter—and so became an opening, an oasis, in the dense fabric of New York City. Though its figural qualities suggest a landscape presence, its openness now suggests an absence.

Perhaps the most famous courtyard garden is Ryoanji in Kyoto, Japan. A visitor enters through the Zen temple and catches a glimpse of this walled "dry landscape" on the left. The focus moves from interior screens and tatami mats to

图 5-01   **Citroen watercolor showing diagonal path cut**
雪铁龙对角线路径切割水彩图

园，每个花园都专门使用一种颜色或风格。这些中等规模的空间为人们提供了可以获得心灵慰藉的场所，是更加内敛的景观，有助于我们专注于内心的平静。银叶屋作为农村和城市景观之间的过渡性媒介，将我们的状态从城市转移到小树林，然后再转移回来。

中央公园是纽约市的一个城市规模的庭院。它在物理空间上用下沉的东西向道路划分了街道，从而创造如画的景色视角，并利用周围建筑物的边缘，在 57 街至 110 街的建筑物之间形成了一个巨大的围合空间。公园内部草地相连，由铺砌和未铺砌的蜿蜒小路分开。水体是其中最大的不间断的空间。中心小路表示道路之间以及两旁树木之间的移动和交换。中央公园见证了周边建筑的发展，并因此成为纽约市密集结构中的一个开口，一个绿洲。虽然

它的形象暗示了一种景观的存在，但它的开放性又代表了一种城市空间的缺失。

日本京都的龙安寺是非常著名的庭院花园。游客通过禅寺进入，可以瞥见左侧这片有围墙的"硬质景观"。观者的焦点从室内的屏风和榻榻米转移到露台台阶外独特的岩石花园。游客坐在低矮屋檐下的台阶上——内部和外部的界限变得模糊了（除非下雨限制了空间流动）。虽然这个精致的花园三面都有墙，而且是封闭的，但它的构成在视觉上无法一目了然。从任何视野良好的位置看，至少有一块岩石是隐藏起来的。砾石中的岩石呈现出海中岛屿的特征。砾石沿北面和东面的边缘由石板铺成，形成了花园和寺庙之间的过渡空间。沿着石板路，蜿蜒的小溪代表鸭川河，以微缩的尺度再现城市的一角。

the distinctive rock garden beyond the terrace steps. Sitting on these steps under a low eave, the boundary between inside and outside is blurred (except when rain limits the use of the outside). Though the refined garden is walled and enclosed on three sides, its composition cannot be seen all at once. From any vantage point, at least one rock is hidden from view. The rocks in gravel take on the characteristics of islands in the sea. The gravel is bound by stone pavers along the north and eastern edges which creates a threshold between the garden and the temple. There is a small brook, a miniature imitation of the Kamogawa River, along these pavers that mimics the section of the city on a much smaller scale.

On a domestic scale, the courtyard gardens of Ryue Nishizawa's Weekend House (2017), Gunma Prefecture, Japan, serve as a bridge to the woodlands beyond. The narrow courtyards change the ambience of the compact interior scale and link it to the larger one.

These three courtyards touch the enclosing skin and are conceived as fragments of the landscape, puncturing

the volume when the doors that separate them from the exterior are opened. Furthermore, they are covered with wood slats that protect them from the sunrays by regulating the intensity of light and are wrapped with glass surfaces whose reflection multiplies on the floors and satiny ceilings, filling the interior spaces with the greenish reflections of the surrounding vegetation.[90]

Our understanding of scale is understood as a relative condition that follows from our own frames of reference. We understand the scale of the environment from our lived experiences. A farmer coming into the city for the first time feels constrained by the density of the built environment. A city dweller going to the countryside for the first time experiences even a small field as expansive. As we age, a season seems shorter; a decade can be contained in a momentary recollection. A familiar landscape—one that we return to often—can yield a sense of coziness, even if it contains an expanse of ocean.

Here we understand the scale of physical conditions as well as temporal ones. We can think of the scale of landscapes in

在私宅规模上，日本群马县西泽立卫的"周末之家"庭院花园充当了通往更远林地的桥梁。狭窄的庭院改变了紧凑的室内尺度，并将其与更大的尺度联系起来。"这三个庭院连接着周围环境，被视为景观的碎片，当将它们与外部隔开的门打开时，景观就会冲出空间。此外，它们还覆盖着木板条，通过调节光线的强度来防止阳光的伤害。包裹的玻璃表面使反射在地板和光滑的天花板上成倍增加，周围植被的绿色充满内部空间。"[90]

我们对尺度的理解是从我们自己的参照系中得出的一个相对条件。我们从生活经验中理解环境的尺度。第一次来到城市的农民会感到被密集的建筑环境所束缚。一个城市居民第一次去乡下，即使看到一小片田地也会觉得广阔。随着我们年龄的增长，一个季节似乎更短；十

年可以包含在一瞬间的回忆中。一个熟悉的风景——我们经常回想起的——可以产生一种舒适亲密的感觉，哪怕它是一望无际的海洋。

我们从这里出发理解物理的规模以及时间的规模。我们同样可以从时间上考虑景观的规模——从地质时期到历史年代，从年度到季节到每日。

美国艺术家丹尼斯·奥本海姆的《时间线》通过一种短暂的状态探索时间尺度。这一作品跨越了加拿大和美国之间的政治边界，沿着一条河流，在冰雪中铲出临时线路。照片显示，作为一项政治行动，这些人在被圣约翰河分割的国家两侧创造了一个半圆形，短暂地改变了河流边缘和政治边界的性质。每个人都挖了一个半圆，深度同积雪的厚度，宽度允许一个人在河边移动。该比例是基于身体决定的，铲的

temporal extent—from geological periods to historical eras, from annual to seasonal to daily.

Timelines by American artist Dennis Oppenheim explores temporal scales through an ephemeral condition. The work straddles the political border between Canada and the United States along a river through the act of making temporary lines in the snow and ice by shoveling. Photographs show the men creating ephemeral semi-circles on either side of each nation divided by the St. John River as a political action, temporarily changing the riverine edge and the nature of a political boundary. Each man has dug a semi-circle as deep as the snow and as wide as permits him to shift around the river. The scale is based upon the body, the act of shoveling, and the act of making a geometric figure visible from a distance through removal.

Another work of absence is Maya Lin's Vietnam Veterans Memorial, National Mall, Washington D.C. This tangible absence provides a place to mourn publicly—through moving along the wall—descending and rising to create "a journey towards an awareness of loss."[91] This landscape's

scale is simultaneously at the human scale (people can touch the polished wall of names), and the larger mall scale (the Lincoln Memorial, and the Washington Monument). This work inscribed on the surface of the public green holds numerous scales of time—the duration of the war and how the span of time after it changes the war's presence in our consciousness. This landscape is formed by juxtaposition and united through a new figure or set of figures. Consider how through this memorial, Maya Lin re-orients two landmarks and puts them into a new conversation. The memorial also connects to the time of the other monuments' constructions, and the time and philosophy of the leaders they recognize (Lincoln and Washington).

A different physical and temporal scale of work created through absence is *Double Negative*, 1969, in Nevada by American artist Michael Heizer. "*Double Negative* consists of two long straight trenches, 30 feet wide and 50 feet deep, cut into the 'tabletop' of Mormon Mesa, displacing 240,000 tons of desert sandstone. The cuts face each other across an indentation in the plateaus' scalloped perimeter, forming a continuous image, a thick linear volume that

动作，就是通过移动使一个几何图形从远处可见的动作。

另一个有关缺席的作品是林璎在华盛顿特区国家广场内设计的越战纪念碑。这种有形的缺席提供了一个公开哀悼的场所——通过沿着墙的移动路线，下降和上升——创造了"一个失去意识的旅程"。[91] 这一景观同时使用了手的尺度（触摸抛光的名字墙）、身体的尺度（回廊和轮式、成人和儿童）和建筑物的尺度（类似林肯纪念馆和华盛顿纪念碑）。这一镌刻在公共绿地表面的作品具有无数的时间尺度——战争的持续时间，以及战争之后的时间跨度如何改变我们意识中战争的存在。如果我们观察图形与地面的关系，会发现这种景观是由相邻区域形成的，通过一个新的图形或一组图形结合在一起。试想一下，林璎是如何通过这座纪念碑重新定位两个地标，并将它们置于新的对话中的。纪念碑还与其他纪念碑的建造时间以及其纪念人物（林肯和华盛顿）的时间有着哲学上的联系。

美国艺术家迈克尔·黑泽尔于 1969 年在内华达州创作的《双重否定》是一个具有不同物理和时间尺度的作品。"《双重否定》由两条长长的壕沟组成，宽约 9 米，深约 15 米，在莫尔蒙梅萨东部边缘的悬崖上凿下了 24 万吨的岩石。这些切口在高原扇形边缘的凹陷处相对而立，形成一个连续的图像，一个厚实的线性体量，这个线性体量连接并结合了它们直接的'负空间'。"[92] 这一工程显然是传世之作。这座雕塑让人想起另一个千年的土方工程——秘鲁纳斯卡巨画的尺度。除了从高空清晰可辨之外，这个项目还包含了工业化进程的

bridges and combines the 'negative' [92] space between them." This work is emphatically not ephemeral. Harking back to earthworks of a different millennium, the sculpture calls up the scale of the Peruvian Nazca lines. In addition to being visible from above, this project incorporates the scale of industrial processes—of excavation through machinery, of intended domination. This work's presence in a mostly uninhabited territory shows the dichotomy of the geological scale and the human.

Another landscape in dialogue with the scale of industrial processes is Peter Latz's Duisburg Nord in Duisburg-Meiderich, Germany, made from the blast furnace plant of Thyssen Krupp Mills. The structures and implied spaces dominate the town. The hulking former gas tank is now a place for scuba training. Here the sublime is present in the latent menace of the steel mill. The scale of the plant dwarfs the scale of the person. The dimensions of the park and conference center are based on the industrial processes and flows. The switching building was transformed into a gargantuan room for a conference (more than 500 feet long and 110 feet) holding more than 4000 people. One senses the prior industrial use in the former concrete storage rooms in the partially enclosed gardens. The sintering plant was demolished, and its toxic landscape is being remediated by flowers in phases—but the underlying organization of the park keeps recalling the scale of heavy manufacturing and history.

Scale is both a physical measure and a relative perception. When landscape architects think of a site's scale, both its actual dimensions and the perception of it in context are considered.

5.3 Experience

Landscape architects use scale to frame experiences—to help signal a sense of shared identity, a shared narrative in outdoor spaces, to reconcile ourselves with our past, present, and future. Manipulating the experience of vertical and horizontal scales, Parc Diderot by Allain Provost is a public park that is dominated by its vertical zones of green and constructed waterfalls (Fig 5-02, 5-03). The human scale is inscribed in the steps and paths. The scale of the city is mediated by a vertiginous garden that connects playing

尺度——通过机械进行挖掘，以及人为的控制。这件作品出现在一个几乎无人居住的地方，在地质尺度和人类之间展开了对话。

另一个与工业化进程规模对话的景观是彼得·拉茨在德国杜伊斯堡-迈德里赫的作品北杜伊斯堡景观公园。该作品由蒂森克虏伯工厂的高炉厂房改建而来。这些结构和隐含的空间主宰着这个小镇。笨重的"煤气罐"现在是潜水训练的场所。在这里，壮丽的景象存在于钢铁厂的潜在威胁之中。工厂的规模和人体尺度形成巨大差异。公园和会议中心的尺度取决于工业厂房的尺度。发电厂被改造成一个巨大的会议厅，约152米长，约33米宽，可容纳4 000多人。在部分封闭的花园里，人们还可以通过混凝土储藏室感受到之前的工业气息。烧结厂已被拆除，花卉正在分阶段修复其破旧

的景观——而公园的底层结构仍在不断唤起对重工业和历史的回忆。

尺度既是物理尺度，也是相对感知。当景观设计师考虑一个场地的尺度时，既要考虑它的实际尺寸，又要考虑它的环境感知。

三、人的体验

景观设计师利用尺度来构建体验——帮助在城市空间中传达一种身份共享感，在户外房间中传达一种共享的叙事，以思考我们的过去、现在和未来。阿兰·普罗沃设计的狄德罗公园是一个公共公园，运用了垂直和水平尺度的经验，以垂直的绿色区域和人工瀑布区为主（图5-02、5-03）。人的尺度体现在台阶和道路上。垂直的花园调和了城市的尺度，将运动场和操场以及上面的住宅连接起来。陡峭的瀑布让人

fields and playgrounds with housing above. The steep waterfall lets one forget about the city and its concrete and lose oneself in a more serene soundscape. The scale of the constructed waterfall parted by vertical paths is compressed so that the waterfall and its stone elements become the visitor's horizon.

This is the nature of contingency—not only bearing on other physical and ecological conditions—but also on personal and emotional ones. The landscape architecture of cemeteries has a special connection to the human scale. Architect Alex Anmahian spoke about the proximity of architecture[93]—how a project changes and should address scale—from 400 feet away[94], from 40 feet away, and from 4 feet away.

In the Woodland Cemetery in Stockholm, Sweden, designed by Erik Gunnar Asplund and Sigurd Lewerentz, one enters along a long high wall that frames a road. As the visitor walks on, the large Woodland Crematorium looms on the left (Fig 5-04).

This classical Nordic building comes into view and hangs over the visitor. The forecourt is a three-story space away from trees—giving it a stark quality that dwarfs the visitor (Fig 5-05). Walking past the crematorium, one must leave the main path to encounter Lewerentz's Woodland Chapel in a grove (Fig 5-06).

This outdoor room in the woods gives way to a threshold space framed by twelve barely dressed timber posts. Stepping beneath this deep eave, one enters the interior chapel. These spaces feel intimate—the chapel is only large enough to hold twenty people. The outdoor room contrasts with the grandeur of the rolling meadow to the west and the monumental crematorium that is encountered first. The heights are the first distinguishing characteristic followed by what is enclosed and what is still the area of the chapel. The 60-foot-tall trees are brought down to the measure of the body through the timbers framing the entry. A grove of trees here gives a sense of the outdoors.

A sapling is analogous to a young body and fallen timbers to one that is buried. Igualada Cemetery by Enric Miralles and

的感官忘记了城市及其混凝土材料，迷失在一个更加宁静的声景中。经过设计后，这座被垂直步道分隔的人工瀑布整体规模缩小了，使得瀑布本身与周围的石头共同形成了游客视野范围内的主体景观。

这就是偶然性的本质，不仅与物理和生态条件有关，也关系着个人和情感条件。墓地的景观建筑与身体的规模和方法有着特殊的联系。建筑师亚历克斯·安曼希恩谈到了建筑的规模性 [93]——项目如何变化，以及如何强调规模——从 400 英尺 [94]，到 40 英尺，到 4 英尺。

在由埃里克·古尔纳·阿斯普朗德和西格德·莱韦伦兹设计的瑞典斯德哥尔摩林地公墓中，人们沿着一堵长长的高墙进入公墓，这堵墙构成了一条道路。随着来访者的深入，大型林地火葬场在左侧若隐若现（图 5-04、5-05）。

接着，一座北欧古典建筑映入眼帘，悬在游客的头上。前院是一个远离树木的三层楼空间，对比之下给游客一种自身渺小的感觉。走过火葬场，人们必须离开主路，在一片小树林中与莱韦伦茨林地教堂相遇（图 5-06、5-07）。

这个树林中的室外房间位于门廊的门槛空间，门廊由 12 根几乎没有修整的木柱构成。走到这个深深的屋檐下，便进入了内部的小教堂。小教堂只够容纳 20 个人，这个空间给人以亲切感。室外的房间与西边起伏的草地和首先遇到的纪念碑式的宏伟火葬场形成鲜明对比。高度是第一个显著的特征，第二个是水平特征——什么是围合的，以及如何界定小教堂的区域？60 英尺高的树木通过入口处的木柱被降到了身体的尺度。这里的树丛给我们一种置身户外的感觉。

Carmen Pinos in Barcelona, Spain, embeds fallen timbers in the ground to remind us of our own physical and temporal measure. This large organizing figure of the Z-shaped plan connects us to a scale of the fields surrounding this work. This project uses descending and rising to create a procession about coping with loss and to remind us of our own time above and below ground. The repetition of smaller precast building elements that can be rotated and fit together with gaps returns the scale of a long wall to an individual. This project's burial walls create an outdoor procession framed by rough concrete with some trees to soften a stark experience. The steep slope down amid a wide passage as we enter is answered by narrow stairs elsewhere that permit us to rise through the walls holding the dead.

When we are not pedestrians, experiencing a landscape while on a train or in a car—the speed changes our perception of the landscape. It becomes a blur. Its larger sweep and scale become apparent. John Brinckerhoff Jackson understood landscape at the speed of a motorcycle. His interest in perception at high speeds alternated with what caught his eye and encouraged him

to slow down and pause. He frequently crossed the country on a motorcycle and made a sketch by the side of the road in the time it took to smoke one cigarette.

A landscape that is designed for approach by an automobile is Reed Hilderbrand's Parrish Art Museum landscape, It announces its presence from the road. The planted fields serve as a foreground for the industrial museum building beyond. As one drives into the parking lot, one encounters a rich landscape that holds spaces in three and four car increments. This grouping breaks down the scale of the lot. These outdoor storerooms lead to wetland plants at the human scale and invite pedestrians to inhale the scent of the plants before crossing over culverts. In this way, the parking lot is held apart from the museum grounds by a zone of water.

Another landscape that plays with cars is City Lounge in St Gallen, Switzerland, by artist Pipilotti Rist and architect Carlos Martinez (Fig 5-08, 5-09). This public project is decidedly of a human scale, welcoming people to jump, lean, rest, and gather in groups. The red surface overtakes

图 5-02　**Park Diderot constructed waterfall**
狄德罗公园建造的作为地平线的瀑布

图 5-03　**Cascading water feature**
模仿瀑布的地形

图 5-04　**Woodland Cemetery entering along low wall, leading to chapels**
高墙延伸进林地公墓，并通往教堂

图 5-05 Woodland Cemetery looking towards the expanse
在林地公墓眺望广袤的空间

图 5-06 Approaching Chapel Lewerentz
接近莱韦伦茨教堂

图 5-07 Roughhewn columns near trees
树木附近有粗糙凿成的柱子

一棵树苗就相当于一个年轻的身体，而倒下的木头就相当于被埋葬的身体。西班牙巴塞罗纳的恩瑞克·米拉莱斯和卡门·皮诺斯设计的伊瓜拉达墓地将倒下的木头嵌入地下，提醒我们关于自己的物理和时间尺度。Z形的大型规划图将我们与作品周围田地的尺度联系起来。作品用下降和上升来创造一个应对逝去的过程，并提醒我们自己在地上和地下的时间。小型的预制建筑构件在这里被重复使用，这些构件可以旋转并通过间隙配合在一起，将长墙的规模还原为个体的规模。这个项目的埋葬墙创造了一列粗糙的户外混凝土架构，并加入一些树木来缓和过于严肃的氛围。当我们进入宽阔的通道时，陡峭的斜坡在其他地方被狭窄的楼梯所代替，让人们可以穿过安放逝者的墙壁。

当我们不是步行，而是在火车或汽车上体验风景时，速度会改变我们对风景的感知，使其细节模糊不清，而规模和轮廓则变得明显。约翰·布林克霍夫·杰克逊以摩托车的速度理解风景。他对高速感知的兴趣与吸引他眼球的东西交替出现，鼓励他放慢脚步并暂时停驻。他经常骑着摩托车穿行一个国家，在路边用抽一支烟的时间画一幅素描。

里德·希尔德布兰德的帕里什艺术博物馆是一个为停车而设计的景观，从道路上体现了它的存在。种植着作物的田地为博物馆的工业建筑提供了前景。当人们开车进入停车场时，会看到一个广阔的场地，宽度可以容纳三到四辆车。这种分组方式分解了停车场的规模。这些室外停车场通向适应人体尺寸的湿地植物区，使行人在穿过涵洞中的水之前可以感受植物的气味。通过这种方式，停车场由一片水域

sidewalk and road to lead visitors to an outdoor room surrounded by buildings with embedded cars as furniture under a rubber blanket. In a city that has original buildings from 900 CE nearby, this decidedly contemporary surface plays off the ancient stone that lines the nearby streets. This surface creates a sense of topological continuity while yielding a discontinuity in time.

A tree canopy up to 2 meters high feels like a natural welcome that many adults can reach with an outstretched arm; to a small child, this same grove of trees may feel like a generous room. We can envision the scale of landscape in a social way: from the measure of a single person, to a small group, to a large group, to a city's whole population. We can conceive of the scale of landscape in a physical way: measurable from a single glance, moving at the pace of walking, cycling, motorcycling, driving, flying. We can imagine the scale of landscape conceptually from the handheld to the domestic, to the civic and regional scales.

Summary

Landscape architecture is always understood through scale. People inherently understand works through perception (visual and acoustic) and movement. This chapter explores exemplary conceptual and physical landscapes to help beginner designers intentionally create conditions perceived in time and space. The case studies demonstrate a range of approaches for people to experience outdoor spaces, passages, and prospects. The examples are drawn from models, art, architecture, residences, playgrounds, cemeteries, and industry across a range of temporal periods.

与博物馆的场地隔开。

另一个与汽车有关的景观是艺术家皮皮洛蒂·里斯特和建筑师卡洛斯·马丁内斯创作的瑞士圣加仑城市酒廊。这个公共项目是为行人打造的，允许人们跳跃、倚靠、休息和聚在一起。红色的表面覆盖了人行道和车道，将人们引向一个室外房间，四周都是建筑物，在橡胶毯下嵌入了汽车作为家具（图5-08、5-09）。在一座建造于公元900年的城市里，极具现代感的表面与附近街道上的古老石头形成鲜明对比。这个表面创造了一种拓扑连续性的感觉，同时产生了时间上的不连续性。

对许多成年人来说，高达两米的树冠是他们伸出手臂就能感受到的令人喜爱的自然，但对小孩子来说，这片树冠可能像一个宽敞的空间。我们可以在社会范围内想象景观的规模：从身体到小群体到大群体，再到城市人口。我们可以在物理范围内想象景观的规模：从一眼就能感知到，再到以步行、自行车、摩托车、汽车、飞行的速度穿行感知。我们可以从概念上想象景观的规模：从手上可以拿着的大小，到一个普通房间大小，到城镇大小，再到地区大小。

小结

尺度是理解景观建筑必要的一环。人们本质上是通过感知（视觉和听觉）和运动来领会作品的。本章探讨了概念性和实体性景观的经典案例，以帮助学生有意识地创造时间和空间的尺度条件。这些案例展示了一系列让人们体验户外房间、通道和前景的方法，体现了不同时期和工业特征，其中包括模型、艺术、建筑、住宅、游乐场、墓地等。

City Lounge St Gallen
图 5-08　圣加仑城市酒廊

City Lounge benches coverd with rubber
图 5-09　覆有橡胶表面的城市休息长凳

课后作业
After-class exercises

讨论题
Discussion

艺术家维哈·塞尔敏通过她的艺术要求我们仔细观察岩石的特殊性。地质学家将岩石与更早的状态联系起来。罗伯特·史密森将真正的岩石带入画廊，并命名了它们的地点起源。这些岩石通过采石、装载和运输的工业过程人为地与它们的初始位置分离，然后以更符合现代艺术画廊的方式展出。

如果在海边拾取一块岩石，就可以通过条纹了解它的一生。我们可以通过岩石中的每一条线画出一幅图，延伸到它的地质层，找到它的矿物含量和它在时间上的位置。人们可以观察到大海对这块岩石一生的记录，可以想象它曾经或者现在属于哪一系统。人们可以将这块岩石本身想象成一个潜在调研基地。在这个微小的尺度上，这块岩石中包含了哪些潜在的景观？

Through her art, Vija Celmins asks us to look carefully at the rock's particularities. A geologist connects the rock to an earlier state.

Robert Smithson brings literal rocks into galleries, and names their site origins. These rocks are disconnected from their place by intention, through an industrial process, of quarrying, loading, and carrying—and then having the rocks conform to the practices of modern art galleries.

If one takes a rock at the seashore, one can see understand it has come into being through striations. One could make a drawing from each line in the rock and extend out to its geological layer, finding its mineral content and its place in time. One can observe how the sea manipulated this rock. One can imagine the systems that it was/is a part of. One can imagine the rock itself as a site. On the miniature scale, what are the potential sites in this one rock?

设计实践
Design practice

请用一块石头思考三个尺度。一个拿在手里，一个需要与普通房间的大小相关，一个视为区域。思考设计师如何在每个尺度中给它注入一个新的概念，请结合图纸进行汇报说明。

Find a rock. Consider three scales: one held in the hand, one connected to the size of a modest room in a house, and one as a region. How does one imbue the rock with a notion of the sublime in each scale?

推荐读物
Reference reading

1. Dee, C. *Form and Fabric in Landscape Architecture: A Visual Introduction*. Taylor & Francis, 2004.
2. Eckbo, G. *Landscapes for Living*. Duell Sloan and Pearce, 1950, especially pp. 61-74.
3. Moscow, K. and Linn, R. S. *Small Scale: Creative Solutions for Better City Living*. Princeton Architectural Press, 2010.
4. Rasmussen, S E. *Experiencing Architecture* (Vol.2). MIT Press, 1964.
5. Simitch, A. and Warke, V. *The Language of Architecture: 26 Principles Every Architect Should Know*. Rockport Publishers, 2014, especially pp. 109-116.
6. Waldheim, C. *The Landscape Urbanism Reader*. Princeton Architectural Press, 2006.

Optional:
1. Bridge, G. "The Hole World: Scales and Spaces of Extraction." *Scenario 05: Extraction*, Fall 2015, viewed online.
2. Vidal, J. M. ed. "Pequeña escala = small scale." Special Issue, *Paisea*, 2014(28).

第六章

景观设计调查与分析方法

Chapter 6
Landscape design survey and evaluation methods

学习要点
Study Points
掌握景观设计调查的方法
Master the method of landscape design survey

学习目的
Teaching Objectives
让学生通过实地考察、视觉评估和体验、资料查阅、信息整理的方法挖掘场地问题，从而归纳出设计问题的核心。
Discover site issues for students through a comprehensive exploration including fieldwork, visual assessment and experience, data review, and information organization to summarize the essence of the design problem.

每一个项目都是从对场地问题的深入探讨开始的。在设计讨论的过程中，不仅需要专业的第一手资料分析，也需要科学的数据以及非专业人士的访谈等二手资料的归纳，从而从多视角提炼问题的核心。本章节从设计过程所涉及的实地考察、视觉评估和体验、资料查阅、信息整理四个方面，为学生展示常规研究的方法和途径，辅助学生在设计过程中找到核心问题，从而进行有效的设计。
Each project commences with an in-depth study into the intricacies of site issues. In order to facilitate design discussions, designers should analyze firsthand information while simultaneously synthesizing secondary information through rigorous scientific data analysis and interviews. It is upon this foundation that the crux of the problem can be distilled from multiple perspectives. This chapter seeks to present to students the methods and approaches of conventional research, focusing on four key aspects intrinsic to the design process (i.e. fieldwork, visual assessment and experience, data review, and information organization). By engaging in these activities, students will be equipped to identify the core design issues and effectively execute their design.

棉 麻
M · M SHO

一、实地调查

1. 实地观察

景观设计对场地自然环境和人居环境的观察，是设计过程中重要的定性研究，目的在于现场数据的记录和统计。[95] 通过观察，可以对场地的设计目的、面对的困难和需要解决的问题有一个快速的认知。

首先是对实地环境的观察。看什么？这是设计师需要思考的问题。从自然环境的角度，设计师需要了解场地及周边是否存在危险或者不利于生存发展的空间。详细来说，可以从自然特征、建筑特征、社会特征三个方面进行观察。[96]

（1）自然特征

气候：景观设计与区域的气候条件有着密切的联系。在场地考察的过程中需要关注周边构筑物或者景观设施是否对气候特征进行了设计回应。例如，在多雨而潮湿的区域，需要有架高的平台避免潮气的蔓延。而对于常年温度高、阳光暴晒时间长的区域，需要有遮阴功能的人行道。我们可以通过观察场地周边现有建筑和构筑物的特征来了解气候对于当地景观设计的影响。

植被带：现场植被的记录观察，主要需要了解植物生长的健康程度和年份，从而确定需要保留的乔木和灌木的范围。成熟的高大乔木具有较高的保留价值，对于形成自然遮阴环境有着天然的优势。随意砍伐有价值的树木，会增加之后培育的成本和时间，并且不利于植被对于土壤的固定。场地植被的生长状态与区域气候、光照、风力和水土条件有密切的关系。观察生长条件有优势的植物类型，有助于设计

6.1 Site analysis
6.1.1 Field observation

In landscape design, adherence to the observation of the natural environment and the human habitat of the site is paramount. A pivotal qualitative survey is conducted with the objective of documenting and quantifying site data. [95] Through field observations, a rapid understanding of the site's layout is attained, thereby unveiling the challenges and predicaments that necessitate consideration in the design process.

By observing the natural environment, designers are able to gain profound insights into the presence of any hazardous or unfavorable spaces in and surrounding the site. To acquire such insights, observations must comprise the characteristics of the natural environment, architectural features, and social attributes.[96] Natural features consist of climatic conditions, vegetation zones, soil composition, water sources, and other pertinent factors. Landscape architectural features include the environmental quality of adjacent parks, recreational areas, sewage facilities, historical and cultural sites, as well as other service facilities. Furthermore, social attributes involve factors pertaining to the construction site, lifestyle, and legal considerations of the land.

Natural features

▪ Climate characteristics: The design of sites is profoundly influenced by the climatic conditions prevalent in the region. During site visits, attention must be paid to whether the surrounding structures or landscape facilities have been designed in response to the climatic characteristics. For instance, in areas characterized by high levels of rainfall and humidity, elevated platforms are necessary to prevent the spread of moisture. Similarly, areas with consistently high temperatures and prolonged sun exposure necessitate the consideration of shaded sidewalks. A comprehensive understanding of the impact of climate on environmental design can be gained by appraising the construction of existing buildings and structures in the vicinity of the site.

▪ Vegetation strips: In general, the acquisition of insight into the health and age of vegetation present on the site is necessary in order to determine the necessity of retaining the range of trees and shrubs. Trees that have

师在选择新植被时获得参考。

土壤和地形：土壤的成分和属性对于场地规划有重要意义。在规划设计中，土壤与土地的稳固性决定地基是否具有可挖掘性、抗侵害能力，以及是否具有下陷风险，还决定了土地的排水状况。同时，土壤的健康程度也会体现在植物的生长表现上。地形是土壤评估中必不可少的一部分。土壤与基础地质的自然侵蚀过程，会形成高低不一的地貌形态。对于山坡、山谷、山脊、丘陵、山地等地形，施工的方法也有所不同。在观察场地的时候，需要准确记录场地本来的地形变化，为土地施工和建筑方案提供基础。[97] 地形的测量可以辅助设计师了解地面排水情况，同时也能显示景色优美的角度，为设计师设计从不同方位观察和测试建筑规划方案的模型提供基础。

野生动物：与植被相关的是野生动物群落。对野生动物的观察需要根据场地植物类型了解野生动物栖息的情况。从昆虫、鸟类和哺乳动物等方面，考虑场地规划和建设是否会影响周边动物的栖息环境。特别是对自然林地的开发和乡村的迁移，在改变了土地性质的情况下，动物的栖息环境就会受到较大的侵害。

水源和支流：场地设计中水源和支流的分布对于项目规划与开发有着重要的意义。支流与地形所形成的景观元素可以丰富场地中的视觉和听觉效果。在前期考察的过程中，要留意水源和支流在场地中的地理关系。在上游区域，场地建造要尽量降低污染的风险。在支流的下游区域，需要评估流量与年降水量的关系，了解场地是否具有洪涝风险。了解场地的水源和地形情况，优化规划模式，可进一步预防洪水

reached maturity and possess considerable height exhibit a significant retention value, thereby reflecting inherent advantages in the establishment of natural shade for the site. The arbitrary felling of valuable tree species leads to an extended duration for subsequent cultivation, thereby impeding the consolidation of the original sap cup in the soil. It is an inescapable reality that the growth state of site vegetation is correlated with regional climatic conditions, luminosity levels, wind patterns, as well as soil and water conditions. By means of observing plant types that demonstrate propitious growth conditions, designers are empowered to make more efficacious choices pertaining to the selection of new vegetation.

• Soils and topography: The composition and properties of soils assume a position of paramount importance in site planning. In the process of planning and design, the soil and land stability serve as determinants of whether the foundation is excavated, resistant to assault, equipped with drainage facilities, and susceptible to subsidence. In addition, the vitality of the soil exerts an influence on the growth of plants.

Topography assumes a pivotal role in soil assessment. Terrain patterns characterized by varying elevations are created through the natural process of soil erosion and the underlying geology. Differences exist between topographic formations (e.g. hillsides, valleys, ridges, hills, and mountains) in terms of their construction approaches. During the course of site observation, changes in the original site topography should be documented to establish a foundation for land construction and building design in the context of site planning.[97] Topographic measurement facilitates the acquisition of comprehensive knowledge pertaining to ground drainage while simultaneously providing a visual perspective. Moreover, it offers the designer a model base that can be observed and tested from diverse angles in relation to architectural planning schemes.

• Wildlife: Wildlife communities are intrinsically linked to plant cover. The observation of wildlife is contingent upon the type of vegetation present on the site. The potential impact of site planning and construction on the habitat of animals in the surrounding environment necessitates consideration. It is worth noting that the habitat of animals

灾害对环境的破坏。

（2）建筑特征

景观设计与建筑设计无法分割，在考察景观条件的同时，建筑环境的优劣也影响着景观设计的方案。其中包括建筑物的使用情况、立面风格、出入口的位置分布，以及相关的市政规划元素，如道路系统的规划、道路等级的划分、排污系统、地下管网、电缆的分布位置等。从而可以进一步了解施工现场以及施工过程中有可能遇到的障碍和问题。[98]

（3）社会特征

对社会因素的考察包括场地建设的法规制度、人类生活方式、大众意愿的合理性等，需要根据国家制定的土地规划标准进行土地规划和项目建设。如用地性质的规范、容积率的限制、道路开发的通行权等问题。并且，对于具

有本土历史和传统文化价值的建筑场所，维护和保留场地的历史遗迹对于区域的未来发展具有重要意义。

另外，场地周边娱乐设施、公园绿地、休闲活动设施的使用状态以及居民的生活流线等问题，社会考察也需要关注。如果在已有城市规划的肌理中进行开发，则需要考虑场地与周边建筑、居民活动流线以及活动习惯等如何形成统一的关系。任何违反本地居民生活意愿和日常生活习惯的设计活动，都会引发公众的抵制。设计师作为综合因素的协调者，需要为不同的场地选择最合适的方案。

2. 互动调查

在场地观察中，除了视觉的观察外，还需要进行互动调查来了解场地使用情况。互动调查可分为访谈调研法和问卷调查法两种类型。

is subject to greater influence with the expansion of natural woodland and the encroachment into rural areas, thereby changing the nature of the land.

• Water sources and tributaries: The distribution of water sources and tributaries in site design is of utmost importance in project development and planning. The preliminary site investigation should pay considerable attention to the geographical relationship between water sources and tributaries on the site. The construction process in the upstream areas must minimize the risk of pollution. The convergence of tributaries and the topography give rise to landscape elements that enhance the visual and auditory aspects of the site. Evaluation of the correlation between the river and annual precipitation in the downstream areas of the tributaries is necessary to determine if the site is prone to flooding. By utilizing the topographic environment and optimizing the planning model, the detrimental effects of flooding on the environment can be prevented or reduced.

Architectural features

The analysis of landscape site conditions necessitates the integration of landscape design and architectural design. The attributes of the built environment (e.g. the use of the building, façade style, the locations of entrances and exits) exhibit a certain effect on the landscape design program. Besides, pertinent municipal planning elements (e.g. the planning of the road system, the division of the road grade, the sewage system, the underground pipe network, the distribution of cables) also impact the landscape design program. Based on these factors, a deeper understanding can be gained regarding the construction process, the road leading to the construction site, and potential obstacles during construction.[98]

Social attributes

Social factors must also be examined, including the rationality of the regulatory system, human lifestyle, and public sentiment towards site construction. The process of site planning requires adherence to state land planning standards (e.g. the regulation of the nature of the site, the restriction of the volume ratio, the right-of-way for road

（1）访谈调研法

访谈调研法根据研究问题的性质、目的和对象进行不同形式的访谈活动[99]，可以分为半结构访谈和结构访谈两种。半结构访谈，主要根据一个简单的访谈提纲进行非正式的访谈。对于访谈问题的回复只需要粗略的表达即可。访谈者在访谈的过程中，可对大纲进行相应的灵活调整，变换提问的方式、顺序和问题。结构访谈则会根据主题内容，层层递进地探讨研究的问题。访问者需要按照问题的顺序和节奏进行访谈，不可以轻易调整。

对于环境问题的粗略普查，访谈调研是非常直接有效的方法。[100]设计师可以与当地的居民和附近的路人进行访谈对话，通过聊天和讨论了解场地潜在的功能隐患和公众生活需求。这种访谈方式的优势在于可以让调查者较为轻松且快速地了解场地使用状况，并通过访谈补充观察不到的用户反馈。同时，访谈调研法可以为设计师开拓思路，根据反馈的内容拓展设计思路。访谈记录的手段是多样的，可以通过文字记录进行归纳，也可以通过现场拍照和短片录制的方式进行图像和音频数据的捕捉。例如，在上海苏州河滨水空间的改造中，学生以苏州河滨水空间为研究对象，针对街道景观特征和步行活动进行访谈，并且根据访谈内容统计分析景观节点的使用情况，并分析景观空间的特征[101]（图6-01）。

（2）问卷调查法

问卷调查法是一种科学转换感受的统计测量方法。通过对特定问题的提炼，统计和展示反馈者的答案，从而获得测量结果。在设计调研中，问卷调查法是常用的环境测评方法，优

development). Moreover, the preservation of architectural sites with local historical significance and traditional cultural values, as well as the protection of historical relics on the site, hold critical importance for the subsequent development of the area.

In addition, several issues necessitate further analysis (e.g. the state of use of recreational facilities, parks, leisure facilities and the residents in the vicinity of the site). If the site development is intended to align with the existing urban plan, consideration must be given to the integration of the site with surrounding buildings and the activities of the residents. Any design activity that deviates from the desires and daily routines of local residents will likely encounter public resistance. In the aforementioned activities, the designer assumes the role of a coordinator, selecting the most suitable solution to address various site-related issues.

6.1.2 Interactive survey
As an art form, landscape design imparts specific forms to landscape materials. The original material elements (e.g. stones, plants and water bodies) serve as the poetic language materials created by the designer. The designer expresses themselves poetically, transforming the landscape into a poem. Based on this premise, various material elements are no longer meaningless combinations of elements but are shaped into landscapes through design.

During the site research process, the designer should record visual symbols to capture the landscape's language features. These features are then combined to create the overall design of the place. The natural language symbols of the site can be depicted through sketches and photography. The landscape language is collected by observing and capturing the elements of the site's landscape. The shape, structure, material, and function of these elements are essential components of the landscape language.

Research interview method
It is worth noting that the research interview methodology comprises two types of interviews[99]: semi-structured and structured. In general, the semi-structured interview approach refers to an informal interview conducted in

势在于采集成本低并且能标准化统计问题测评结果。问卷调研法可以通过现场访问或网络投票的方式进行（图 6-02）。但在制作调研问卷的过程中，需要有清晰的问题导向，提问需要与项目内容相关，避免广而泛的回复，导致测评结果没有设计导向意义。问卷调查法虽然普遍适用于各种场地，但是劣势在于提问者主观设定的问题和数据的统计内容具有局限性。问卷调查法并不能简单地解决问题，设计师需要将采访数据与现场了解到的真实情况相结合，进行比较及核对，从而找到适合场地发展的途径。

（3）数据记录

场地调查通过整合文字资料和现场统计的数据，从而对调研资料进行审核、比对、讨论、分类和汇总。[102] 文字资料可以通过互联网上的国家地理数据库、地区文化历史检索、人口普查数据等渠道收集。现场数据则是设计师在场地调研时通过图像采集、现场测量、访问统计等方式收集的第一手资料。

地图信息处理可以参考以下网站。

谷歌地球。谷歌地球显示平面及三维图像信息，并且可以下载模型和整理图像（图 6-03）。

百度地图开放平台。从百度地图开放平台可以获取实时交通、用户使用密度、面积指标、建筑属性等信息（图 6-04）。

地理信息系统（GIS）。地理信息系统是以地理空间数据为基础的地理模型分析方法，可以提供地区地质灾害、生态系统、文化遗产、休闲旅游等信息（图 6-05）。

accordance with a basic interview framework. In addition, the responses to the interview questions should be expressed in a general manner. Based on this foundation, the interviewer possesses the flexibility to adapt the interview process (e.g., the manner of posing questions, the sequence and content of the questions) during the course of the interview. Conversely, the structured interview method involves a systematic exploration of the research questions, building upon the primary question. The interviewer must adhere to the sequence and rhythm of the questions, refraining from changing the content of the interview.

Interviews are regarded as a direct and efficacious means of conducting a preliminary assessment of environmental concerns.[100] During site surveys, designers have the opportunity to engage in interviews and conversations with local residents and passers-by, thereby gaining insights into and discussing potential functional hazards and public requirements associated with the site. The interview method offers the advantage of enabling investigators to promptly understand the use of the site, with feedback being supplemented through the interview process.

Moreover, the research interview method can foster innovative ideas that enable designers to expand their design concepts. Interviews can be documented through various means (e.g., summarizing questions via written records, capturing visual data through on-site photography and video recording). For instance, in the renovation of the Suzhou River waterfront space in Shanghai, interviews were conducted to explore the characteristics of the street landscape and the activities taking place in the area. The utilization of the space was analyzed through interviews with individuals present in the vicinity. Furthermore, the characteristics of the landscape were assessed[101] (Fig 6-01).

Questionnaire method
The questionnaire method pertains to a statistical measure whereby results are derived by refining specific questions and statistically presenting the responses. In design research, the questionnaire method has been widely embraced for the analysis of environments. This method offers several advantages, including cost-effectiveness in data collection and the ability to standardize results. However, it is crucial to establish a clear purpose at the

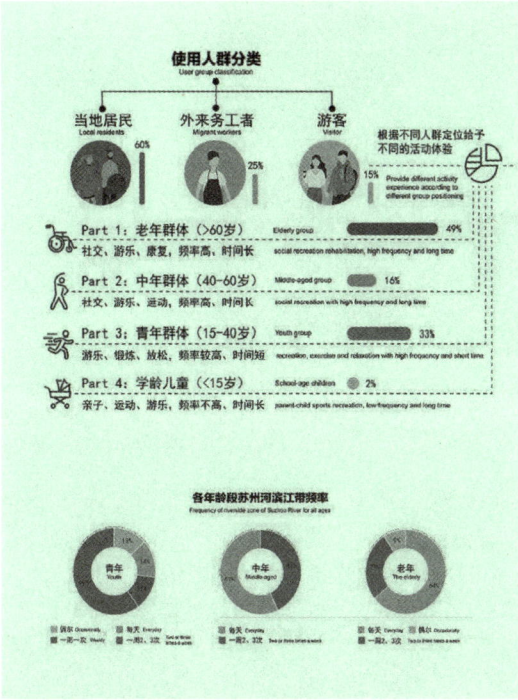

图 6-01 Suzhou River user analysis survey
苏州河滨水空间用户分析调查

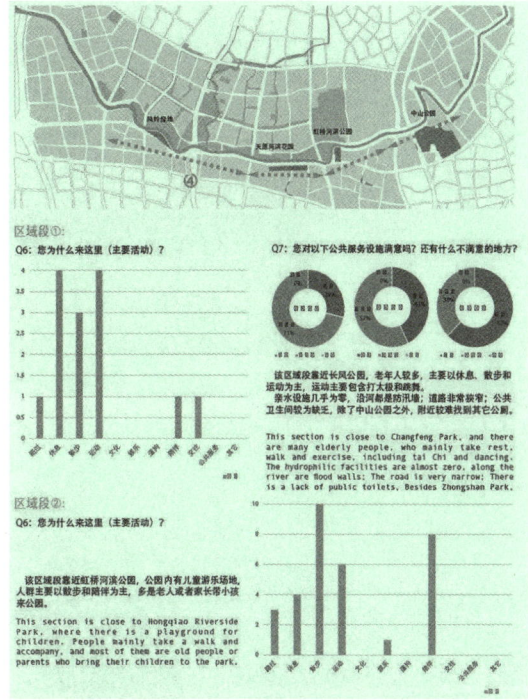

图 6-02 Suzhou River user survey statistical questionnaire
苏州河滨水空间用户调查问卷

core of the questionnaire creation process, ensuring that the questions pertaining to the site are aligned with the intended purpose to avoid broad and generic responses. The assessment results are not intended to guide the interpretation. Despite the questionnaire method's general applicability to diverse site evaluations, its limitation lies in the subjective nature of the data's statistical content and the answers formulated by the questioner. Diverging from mere problem-solving, design problems necessitate designers to compare and cross-reference the interview data with the actual site conditions, thereby identifying the most suitable developmental solution (Fig 6-02).

Data recording

The site survey integrates textual data and field statistics, facilitating the review, comparison, discussion, classification, and summarization of research information.[102] Textual data can be obtained through online exploration of national geographic databases, regional cultural history, and census data. On-site data refers to firsthand information collected by designers through image acquisition, on-site measurement, online

Geographic Information Systems (GIS), and on-site interview statistics.

For map information processing, the following websites can be consulted:

▪ Google Earth: This platform presents both flat and 3D image information, allowing for the downloading of models to organize images (Fig 6-03).

▪ Baidu Map: This resource combines real-time traffic information, user usage density, area indicators, building attributes, and other relevant data (Fig 6-04).

▪ Geographic Information System (GIS): This method of geographic model analysis, based on geospatial data, provides information on regional geological hazards, ecosystems, cultural heritage, and leisure and tourism resources (Fig 6-05).

6.2 Visual assessment and experience
6.2.1 Landscape visual symbols

As an art form, landscape design imparts specific forms to landscape materials. The original material elements (e.g. stones, plants and water bodies) serve as the poetic

图 6-03 **Google Earth**
谷歌地球

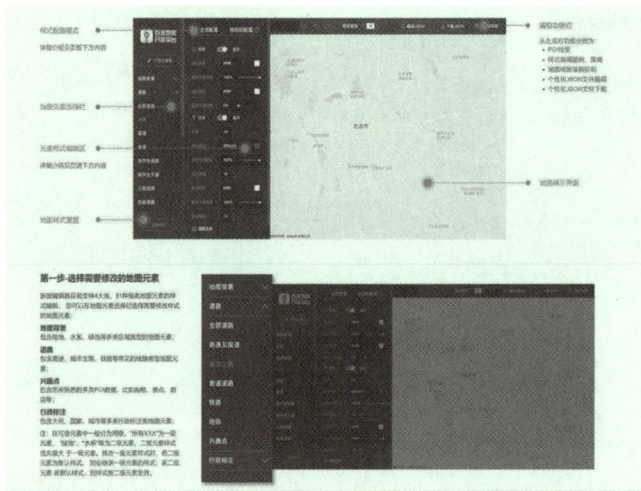

图 6-04 **Baidu Map Open Platform**
百度地图开放平台

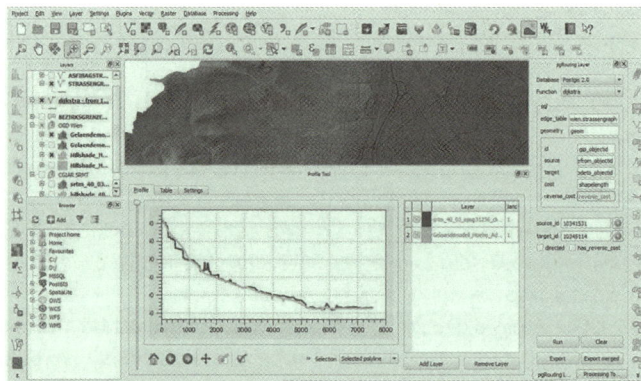

图 6-05 **Geographic Information Systems**
地理信息系统

二、视觉评估和环境体验

在景观设计场地调查的过程中，常常需要通过捕捉场地内的景观语言符号、生态形态、建筑结构、材料等来进行设计的前期分析。本小节根据主题内容，结合实际的调研问题，对调研过程中的语言符号和体验记录进行具体的介绍。

1. 景观的语言符号

"作为艺术，景观设计把一定的形式赋予景观材料，原本不会说话的石头、植物、水体等物质元素就成了设计师诗意创造的语言材料，设计师诗意地言说，景观就成了诗，各种物质材料就不再是一些元素的无意义组合，它们通过设计获得景观的形式。" [103]

场地调研时，设计师需要通过记录视觉符号去收集场地中材料和环境构成的景观语言特征。通过语言特征的提炼，组合成新的空间语言，从而形成场所设计的元素。可以通过速写、摄影的方式对场地自然语言进行描绘。尽可能观察和捕捉景观元素，收集可以激发设计思考的语言。其中，形状、结构、材料、感官元素都是景观语言的重要组成部分。

（1）形状

自然界的形状千变万化，从自然中提取几何形状转化为设计元素是景观设计常用的手法。在中世纪的波斯，出于对数学几何元素的热爱，地毯艺术采用几何图案进行界面的分割，通过几何形状的穿插、组合、缩放等创造新的视觉体验。这种几何形状的应用，被法国园林转化为经典的"结纹园"[104]（图 6-06）。在当代的景观设计语言中，几何形状作为工业化建造技术的表现，被应用在多种地方。例如，形态

language materials created by the designer. The designer expresses themselves poetically, transforming the landscape into a poem. Based on this premise, various material elements are no longer meaningless combinations of elements but are shaped into landscapes through design. [103]

During the site research process, the designer should record visual symbols to capture the landscape's language features. These features are then combined to create the overall design of the place. The natural language symbols of the site can be depicted through sketches and photography. The landscape language is collected by observing and capturing the elements of the site's landscape. The shape, structure, material, and senses are essential components of the landscape language.

Shape

Nature exhibits ever-changing forms, and the transition from natural elements to design elements has been extensively utilized in landscape design. Medieval Persians, who had an affinity for mathematical geometric elements, incorporated these elements into carpet design. By interweaving, combining, and scaling geometric shapes, they created novel visual experiences. French garden designers transformed this application of geometric shapes into the classic "knot garden"[104] (Fig 6-06).

In the field of contemporary landscape design, geometric shapes serve as an industrial construction technique for various landscape elements. For instance, the quantitative production enabled by geometric shapes can be applied to different types of paving, large seating designs, and walls that adapt to the site's topography. The design form of the geometric matrix in the Jewish Museum Square in Berlin, Germany, enhances the site's serious and solemn atmosphere while blending with the surrounding modern architecture (Fig 6-07).

Moreover, the clever transformation of visual shapes can serve as a catalyst for the generation of diverse design concepts. In an urban revitalization project in Shanghai, the design team observed the shape attributes of residents' drying racks on a daily basis, subsequently ingeniously repurposing them into a porch system. This

各异的铺装、可复制的多变座椅、根据场地地形变化的景墙，都利用了几何形状带来的量化生产便利性。德国柏林犹太纪念广场的几何矩阵设计，增强了严肃庄重的气氛，但又与周边的现代建筑语言融为一体（图6-07）。

同时，巧妙转化生活场景中的视觉形状，也可激发设计营造的各种可能。在上海的城市更新项目中，设计团队通过观察居民日常生活中晾衣架的形状特征，设计了廊架系统。廊架可提供遮阴，也可以用来晾晒衣服（图6-08）。

（2）结构

了解场地现有建筑结构，有助于制定能耗节约并且结构稳定的设计方案。现有建筑和建造基础对场地的适应性较强，并且有较强的抗风险能力。借鉴现有结构形态进行空间更新，有助于设计师传递新旧肌理。如上海黄浦滨江

的空间设计，结合旧有码头轨道进行铺装，通过保留原有轨道的结构形式传达历史场景的转换。法国圣殿广场的设计[105]，通过保留19世纪的金属装饰柱网，形成通透的玻璃建筑结构（图6-09），定义了新的广场形式，同时设计元素中对历史文化的显示让新的空间场所更具有时空交错的韵味。

（3）材料

对景观材料的观察可从属性上进行分类。自然属性的材料可以通过观察自然环境获得，如植物层次会随着肌理材料的不同而产生丰富的变化。树木的躯干肌理、枝条的生长结构，以及叶片的颜色质感都会对空间场所的体验具有定型的作用。在步行体验上，沙粒、泥土、木头、草地、石头、混凝土铺装等传达着不同的步行体验。对于休闲场景，可使用更为亲近

innovative approach also encourages residents to utilize the racks for both shading and clothes drying purposes (Fig 6-08).

Structure

A comprehensive understanding of the structural configuration of preexisting buildings in the designated area facilitates the development of energy-efficient and structurally sound design solutions. The existing buildings and foundations effectively adapt to the site, thereby enabling designers to seamlessly integrate the old and new textures. For instance, the Huangpu riverfront design in Shanghai integrates the paving design with the original dock tracks, thereby effectively conveying the transformation of historical scenes. Similarly, in the Place du Temple in France, a permeable glass structure was developed, while retaining the nineteenth-century decorative metal column network [105] (Fig 6-09). The form of new square is defined, while the new area is endowed with a more temporally intertwined ambiance through the exhibition of historical design elements.

Materials

Landscape materials can be systematically classified based on their inherent properties. Materials that exhibit natural properties can be identified through observations of the natural environment. For instance, the variation in plant layers contributes to a rich tapestry of texture materials. The texture of tree trunks, the growth structure of branches, and the coloration of leaves collectively define the experiential qualities of a place. The choice of walking surface, be it sand, soil, wood, grass, stone, or concrete pavement, significantly influences the experiential aspect and allows for varying walking speeds. Therefore, one can leisurely traverse the natural environment and immerse oneself in the captivating landscape.

In the context of urbanized office settings, the combination of artificial materials assumes paramount importance. Concrete, glass, stainless steel, steel plates, and acrylic materials are commonly employed in the creation of urban landscapes. The impact of a material on the landscape can be understood through analysis of its texture and the relationship between different materials.

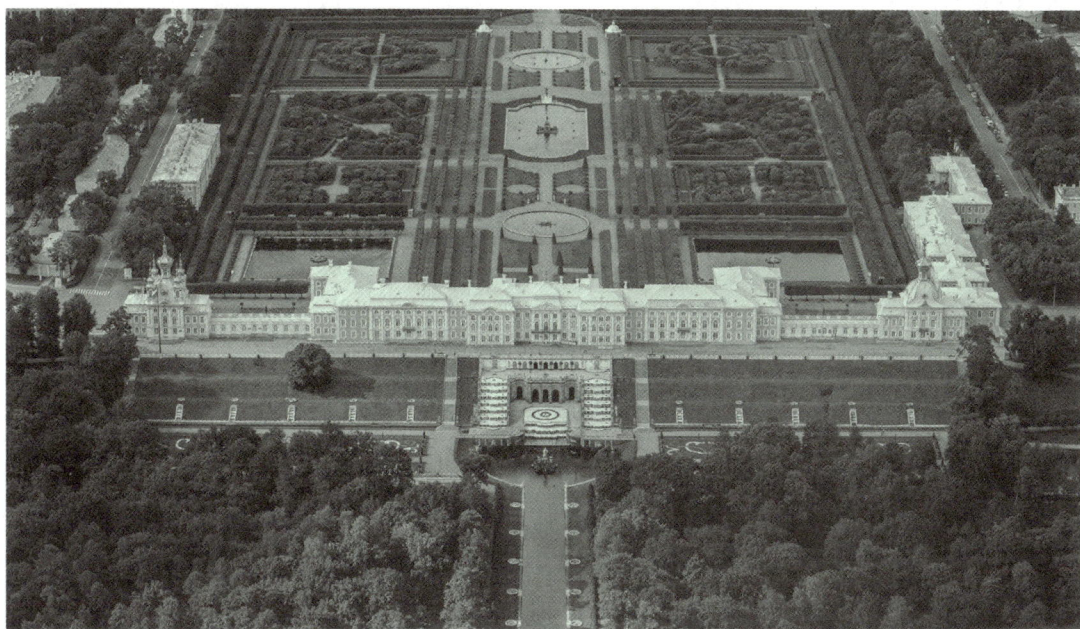

Knot garden
图 6-06　结纹园

Berlin Jewish Memorial Square
图 6-07　柏林犹太纪念广场

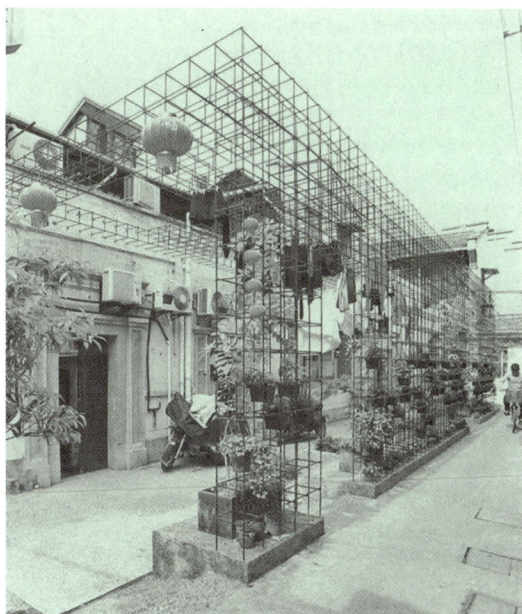

Outdoor trellis
图 6-08　廊架系统

自然的材料组合，让体验者在游走的过程中感受自然。而对于都市化的办公场景，则可以使用人工材料组合，混凝土、玻璃、不锈钢、钢板、亚克力等都是常用的营造材料。通过材料质感和组合变化，我们可以了解材料所传达的肌理效果以及空间个性，从而为不同类型的景观空间配置合适的材料。而同一种材料也会因为加工形式的不同，形成不一样的空间气质。例如，传统日本园林石头的造景方式与经过打磨的石头雕塑会形成自然和抽象两种氛围（图6-10、6-11）。所以，设计师需要不断探索和尝试材料的组合和加工，从而形成有创意的空间场景。

（4）感官元素

景观空间是由建筑和自然环境共同组成的，同时也少不了丰富场景变化体验的活动设计。在场地调研中，视觉、听觉、触觉、嗅觉、味觉这五感是可以观察的感官元素。

视觉：视觉体验是景观空间中最为直观的设计元素，可从色彩、质感、图形这三个方面进行观察。人的感官对色彩的刺激敏感，色彩浓烈的光源所形成的图像令人印象较为深刻，而冷色系的光源视觉影响降低。颜色的冷暖色差会改变人们对于环境空间的感受。[106] 环境中的暖色系光源带来和谐、安宁、温暖的心理感受，冷色系光源带来高贵、神圣、纯粹的心理感受。通过观察环境色彩，可以对场地环境的心理基调形成基本的认知。质感也是视觉中较为明显的元素，可以观察植物、石头、山体、水体的天然质感，也可以观察玻璃、塑料、钢材等人工材料带来的视觉变化。视觉体验中的图形记录，可以被理解为建筑和自然环境中几何抽象元素的构成形态。笔直的高速公路或者

Hence, it is necessary to judiciously select materials that coordinate with the landscape space. In addition, it is worth noting that the same material can evoke different effects contingent upon its form. For instance, traditional Japanese garden stone landscaping and polished stone sculptures evoke two atmospheres, one rooted in nature and the other in abstraction (Fig 6-10, Fig 6-11). Accordingly, the suitability of materials necessitates constant exploration and experimentation to form innovative and captivating environments.

Senses

The landscape space is a composition of architecture and the natural environment, and its efficacy is contingent upon how the space is utilized. In the process of site research, a holistic understanding of the landscape experience is achieved by engaging all five senses.

• Sight: The visual experience, as an inherent aspect of landscape spaces, assumes a paramount role in design. It comprises three fundamental components: color, texture, and graphics. The chromatic palette of the environment stimulates the human sensory apparatus, with warm hues leaving a more indelible impression, while cool tones mitigate the visual impact. Moreover, the difference between warm and cold colors results in differentiated perceptions of the environmental space.[106] Warm light imbues the surroundings with a sense of harmony and tranquility, whereas cold light evokes notions of nobility, sanctity, and fragmentation. By observing the environmental color, one can acquire a rudimentary understanding of the tonal characteristics of the site. Texture, a more significant visual element, manifests itself in the inherent qualities of flora, stones, mountains, and bodies of water, as well as in variations in texture. In artificial environments, visual transformations attributable to diverse textures (e.g., glass, plastic, and steel) can also be observed. Architectural and natural compositions can be distilled into abstract geometric elements. A straight highway or a meandering mountain path can be transmuted into visual graphic elements. Moreover, graphic elements can generate psychological suggestions pertaining to the environment. Circular forms foster cohesion and interaction, while triangular configurations manifest as sharp, conflicting, and dynamic spaces.

图 6-09 **Place Templar, France**
法国圣殿广场

图 6-10 **Garden sculpture, Isamu Noguchi**
野口勇的造园雕塑

图 6-11 **Traditional Japanese garden stones**
传统的日本园林石

蜿蜒的山间小路都可以转换为图形记录。同时，图形元素也会带来环境的心理暗示，如圆形多表达为结合和互动，而三角形多表达为尖锐、冲突、动态。

听觉：声音环境的设计已经成为当代设计中重要的元素。其中，根据健康声音环境的评定，景观设计需要协调场地中的交通噪声、高分贝的设备噪声和街道噪声。[107]合理的规划设计可以在前期改善具有较高噪声的功能区块，将其与居住区分开。另外，也可以通过植物空间配置，将不同活动气氛的空间分类，避免相互打扰。同时，景观设计中对于声音的研究已经不局限于噪声的控制。声景学提出通过对声音环境生态进行针对性的设计，营造健康愉悦的人居环境。环境中的声音被看作设计元素的一种，设置在空间体验的环节中。如自然生境植物设计通过植物的营造，吸引丰富的鸟类群落，从而形成模仿自然的声音效果（图6-12）。

另外，声音的波动频率也可以成为评价环境宜居舒适度的标准。通过在公共环境中记录日常生活噪声的指数，分析具有噪声危害的场景，从而关注人居环境中的声音污染情况。

触觉：触觉主要发生在具有不同材料的体验场景。肢体的互动可以拉近与自然环境的关系。例如，加拿大多伦多的糖果海滩[108]通过在市区中营造沙滩的场景，让周边的工作人员可以从穿着正式服装的办公状态转移到可以脱下鞋去体验细沙的户外状态（图6-13）。设计带来的休闲空间变化通过触觉表达，改变了人们的体验。另外，表现为湖面、雾气、喷泉等形式的水景元素，也为丰富自然景观场所提供触觉（图6-14）。我们可以通过在场地调

• Sound: The auditory environment has emerged as a pivotal facet in contemporary design in the context of site studies. Notably, landscape design should coordinate with traffic noise, high-decibel disturbances, and street clamor to foster a salubrious soundscape[107]. By addressing noise-generating functional zones at the outset and segregating them from residential areas through planning and design, mutual disturbances can be averted. In addition, the scope of sound research in landscape design has transcended the mere control of noise. Deliberate soundscape design can result in a wholesome and pleasurable human habitat. Sound is regarded as an integral design element and an experiential component. As illustrated in Figure 6-12, landscape design, created by the presence of plants, attracts a diverse avian community, thereby fostering a natural acoustic environment.

In addition, the frequency of sound fluctuations can serve as a metric for evaluating the livability and comfort of the environment. By documenting the index of daily life noise in public spaces and analyzing the infringement of noise pollution, greater attention can be directed towards mitigating the encroachment of auditory disturbances on the habitat.

• Touch: The relationship between humans and the natural environment is strengthened through physical interaction, as exemplified by the Sugar Beach in Toronto[108], Canada (see Figure 6-13). In this instance, the personnel in the vicinity transition from formal attire in the office setting to a barefoot experience, simulating a beach scene in the urban landscape. The introduction of fine sand changes the tactile sensation, thereby influencing the emotional state of the staff and creating a recreational space. Additionally, water features such as lakes, fog, and fountains (see Figure 6-14) are incorporated to enhance the tactile elements of the natural surroundings. Thus, the site research reveals the presence of adaptable materials and the potential for transforming the spatial experience.

• Smell: The identification of unpleasant odors is a crucial aspect of site research. The enhancement of olfactory conditions is achieved through re-planning and ecological modifications. In landscape design, olfactory considerations

Natural habitat garden, Chicago, USA

图 6-12　美国芝加哥自然生境植物设计

Sugar Beach, Canada

图 6-13　加拿大糖果海滩

研中观察潜在的可转换材料，思考不同空间需求下触觉体验的变化。

嗅觉：场地调研中，通过嗅觉视角可以分辨出具有不良气味影响的环境因素，并通过重新规划和生态环境改造改善气味环境。嗅觉设计的另一方面，是根据不同的场景进行植物香味的营造。具有香味的植物对于人们感受环境特质是非常重要的。在中国园林中，观赏性的庭院经常会种植丁香花或者紫藤花作为点景的植物，其香气也调节了庭院的气氛，让感官真正地放松。同时，香景学的提出标志着自然香景与人文香景的研究进展。气味要素在不同用途、地域、时间上对于人们感知环境、历史、文化有着重要的影响。[109]

味觉：景观设计中的味觉体验，主要体现在果实蔬菜的种植中。增加可食用的瓜果蔬菜

类配置，可以丰富场地的体验感受（图6-15）。

三、资料查阅

在完成基地的现场调查后，需要结合资料的查阅完善对调研的概括分析。本小节通过案例分析与主题检索、非专业人士调查两个方面介绍查阅资料的方法。

1. 案例分析与主题检索

设计调研主要针对文字资料和数据资料进行整理，形成项目场地设计的基本框架。其中，案例分析是最直接的调查方法，让设计师突破地域的局限，找到相似的设计案例进行横向分析。

场地设计的复杂程度具有很强的地域特征，与项目的目标、面积、性质、设计团队的组合都有密切的关联。虽然国内外项目场地设

involve the cultivation of fragrant plants in various scenarios. The inclusion of plants with distinctive scents contributes to the overall environmental experience. In Chinese gardens, for instance, lilacs or wisteria are often planted in ornamental courtyards as accent plants, imbuing the space with a fragrant ambiance and providing a soothing sensory encounter. Furthermore, ongoing research explores the olfactory landscapes of both nature and human culture, assessing the impact of scents on individuals' perception of the environment, history, and regional characteristics.[109]

• Taste: The taste experience in landscape design is primarily reflected through the cultivation of fruits and vegetables. By incorporating edible plants into the design, the site experience is enriched (Fig 6-15).

6.3 Data research
6.3.1 Case studies and research
Design research predominantly relies on textual and digital information to establish a fundamental framework for addressing design issues specific to the project site. Specifically, case studies have been identified as the

most direct investigative method, enabling designers to transcend geographical limitations and apply analyses of similar design problems.

The complexity of site design problems is closely related to geographical factors and exhibits a significant correlation with the project's objectives, scale, nature, and composition of the design team. Despite variations in the specific challenges encountered in domestic and international projects, the case study approach allows for the application of core design strategies across different design contexts of the same nature. While the design contents may differ among projects, they can all be addressed through similar design strategies.[110] The utilization of case studies facilitates designers' rapid comprehension of site-specific issues, design codes, and applicable standards for a given project type. Case studies can be employed as both project typologies and research methodologies.

Search by project attributes
The collection and retrieval of case studies utilize the attributes inherent in landscape design project properties.

计具体的问题表现都是不同的，但是案例分析的方法就是通过同类型不同设计背景的案例提炼核心设计策略。即使项目设计内容各有不同，也可以通过同类设计策略参考相似的做法。案例分析的方法可以帮助设计师在短时间内了解特定项目类型需要面对的场地问题、设计规范以及标准的具体表达。[110] 案例分析可以从项目属性、研究主题和方法、时间维度三个方面进行检索和搜集。

（1）按照项目属性进行检索

按照景观设计项目的性质进行案例搜索，查看不同国家、不同尺度、不同建造环境、不同投资预算等条件下同类型项目的设计方法。景观设计按项目属性分类可以分为以下类型：森林公园和林地景观、公园、广场、街道景观、滨水景观、体育娱乐空间、居住区景观、商业景观、屋顶花园、居住区景观、工业景观空间、文化遗址保护景观空间等。

也可以根据尺度范围进行分类搜索，可分为区域规划尺度、空间设计尺度、细部设计尺度等，关注不同尺度下景观设计的规范和标准。

还可以从功能和目标方面进行搜索。面对城市多元的发展需求，城市景观项目也从单一的功能开发转向多元发展。以滨水空间为例，城市中的滨水空间设计不仅需要考虑水循环环境的生态设计，也需要关注景观空间对于商业发展价值的提升（图6-16）。在滨水漫步的体验中，增加儿童娱乐设施或者亲水空间来拉近居民与自然环境的距离，可以增强社区的幸福感（图6-17）。所以，面对城市地块的综合开发，设计师可以根据项目愿景进行景观功能的规划，并了解不同功能的设计方法，从而

图 6-14 **Tanner Fountain**
唐纳喷泉

图 6-15 **Urban farm**
城市农场

Chicago riverwalk

图 6-16　芝加哥滨河步道

Shanghai Baoshan riverside

图 6-17　上海宝山滨江

在设计规划初期探索多元可能。

国内外案例搜索网站参考如下。

Archidaily：https://www.archdaily.com/

WLA：https://worldlandscapearchitect.com/

Landzine：https://landezine.com/

Archinet：https://archinect.com/

LAND 8：https://land8.com/

ASLA：https://www.asla.org/

Dedzeen：https://www.dezeen.com/

Landscape Institute：https://www.landscapeinstitute.org/

谷德设计网：https://www.gooood.cn

景观中国：http://www.landscape.cn

（2）按照研究主题和方法进行检索

大部分设计项目会根据设计类型进行项目的定位，也有部分设计项目是根据设计研究的规划进行定位的。设计研究内容具有前沿探索性，通常与国际设计问题和国家的国土空间规划具有密切的关系。例如，以生态文明为主题进行的景观设计研究，主要有海绵城市、弹性城市、可持续城市、低碳城市等设计方向；针对特殊人群进行的景观设计规划，可分为适老化设计、儿童友好社区、医疗景观空间、无障碍设计等；以数字技术为目标进行的元宇宙空间体验、智慧城市、智慧公园等主题研究。另外，以研究工具为对象，也可以进行案例的分析检索，从而了解同种方法对应不同案例的做法。例如，通过地理信息系统、城市兴趣点（POI）、城市热力图等图像模型进行场地问题的分析，并通过地理信息的数据进行问题的剖析。[111]

（3）按照时间维度进行检索

景观设计也可以从时间维度进行案例的归

The design methodologies employed in similar project types can be classified based on landscape attributes and analyzed in the context of various countries, scales, construction conditions, investment budgets, and other relevant factors. Landscape design can be categorized into the following types based on project attributes: forest park and woodland landscape, park, square, street landscape, waterfront landscape, sports and entertainment space, residential landscape, commercial landscape, rooftop garden, residential landscape, industrial landscape space, as well as cultural heritage preservation landscape space.

The search process can also be classified according to scale, which comprises regional planning, space design, and detailed design for the functional exploration of project design details. Irrespective of scale, the development of landscape design should adhere to established design norms and standards.

The evolution of urban landscape projects has transitioned from singular functional development to including multiple demands associated with urban progress. For instance, the design of waterfront spaces in cities necessitates consideration of ecological design principles for the water environment, as well as the value enhancement of land for commercial development (Fig 6-16).

On the waterfront, pathways, children's recreational facilities, or water-friendly spaces are incorporated to foster a closer connection between residents and the natural environment, thereby enhancing the community's sense of well-being (Fig 6-17).

Therefore, designers confronted with the integrated development of urban plots can strategize landscape design in alignment with the project vision, while also gaining insights into the design methodologies pertaining to different landscape functions, thereby exploring a multitude of possibilities during the early stages.

Domestic and international case search site reference
· Archidaily, https://www.archdaily.com/
· WLA , https://worldlandscapearchitect.com/
· Landzine, https://landezine.com/

纳。参考世界景观设计发展史，比较景观设计发展不同时期对于设计问题的思考，从而理解设计方法的形成与社会时代、技术发展、经济基础、人文特征的联系。通过历史的角度归纳设计项目的特征，了解景观设计发展历程中的设计规律，以更长远的视角思考未来景观空间的需求（图 6-18）。

设计师可以从不同的调查角度进行案例的搜索，通过转换视角、检验研究方法、探索不同主题、了解发展历史，形成多角度和深层次的调研学习。

2. 非专业人士调查

在场地调研的过程中，除了专业设计人士对于设计项目的参与，还需要非专业人士的介入，才能更为全面、完整地解决设计问题。很多设计项目从初期到落地都没有公众的参与，形成了"曲高和寡"的现实状况。设计师的初衷总是美好的，希望体验者可以从美学和功能的角度享受更优质的生活。可是，我们也容易局限在自己的思维里，以自身为标尺去衡量别人。所以，在设计的过程中，需要第三方非专业人士的介入，让设计规划满足更多人群的需求。

其中，访谈调研、问卷调查、公众参与式活动都是景观设计常用的第三方介入途径。在前面的章节，我们已经对这三种调查方式进行了介绍。从设计进程角度出发，我们可以把问卷调查设定在不同的项目进程中，从而形成定期审视的流程。例如，在前期的设计访问、阶段性设计汇报、设计使用反馈中都可以通过问卷调查快速了解体验者的情况和需求。访谈调

· Archinet, https://archinect.com/
· LAND 8, https://land8.com/
· ASLA, https://www.asla.org/
· Dedzeen, https://www.dezeen.com/
· Landscape Institute, https://www.landscapeinstitute.org/
· Goode Design, https://www.gooood.cn
· Arch China, http://www.archina.com
· Landscape China, http://www.landscape.cn

Search by research theme and methodology
Most design projects classify the project based on design type, while some are categorized according to design research planning. Specifically, design research content is characterized by its cutting-edge and exploratory nature, often closely intertwined with international design issues and the urban planning of the respective country. For instance, landscape design research can be conducted in environmental design (e.g., sustainable, low-carbon). Landscape design planning targeting specific demographics can be tailored towards age-friendly design, child-friendly communities, medical spaces, barrier-free design, and so forth. In addition, themed research (e.g.,

metaverse space experience, smart city, and smart park with digital technology) is positioned as the objective. By utilizing research tools as the subject, case study searches can be conducted to understand the design methodologies associated with similar design types. For instance, site issues can be analyzed using GIS (geographic information system), POI (points of interest), urban heat maps, and other image models. Additionally, problems can be dissected based on geographical factors.[111]

Designers can conduct case study analyses from various investigative perspectives, while considering different research methods and exploring thematic elements, thereby facilitating multi-faceted and in-depth research.

Case study by time
From the perspective of global development throughout history, the correlation between landscape design methodologies and the social era, technological advancements, economic foundations, and human characteristics can be assessed. In addition, an analysis of design projects from a historical standpoint can provide

研和参与式活动则多举办在设计的前期，因为这两种调查方法都需要花费较长的时间与调查对象进行深入的访谈，通过互相对话的启发产生有利于设计决策的想法。设计前期需要这样大量的多视角问题进行方案孵化，而设计后期则需要更直接明确的决策来指导项目的落地。

四、信息整理

信息整理指对设计调研内容进行集中分析，以数字化的方式展示调研结果。一个图像学的名词也常常出现在设计调研的课堂上，英文是"Mapping"，中文的学术表达为地图术，也称地图学。[112]地图术就是对前期资料进行审查、检验、分类、汇总的全过程（图6-19）。

景观设计运用地图术的目的是将地图和调研的具体数据通过有效的信息图像进行设计问题的表达。设计分析是理性和感性过程的结合，单纯依靠数据进行设计判断会导致设计决策过于生硬，缺乏人文社会关怀，无法合理地反映公众的需求。图像学的表达方法，为设计师提供探索不同创作手法的途径，从而多元地解决问题。例如，以场地的历史文脉为基础，通过历史场景的时间轴展现城市更新的空间形态，利用地图术，对建筑立面、铺装肌理、休闲设施布局等从营造元素的视角进行场景与人活动关系的探索（图6-20）。设计类研究文献也常常通过具体时间案例来说明设计师探索出那个年代的问题和常规途径。本章在最后参考文献部分列出两篇关于设计调研跟踪的文献，可参考拓展数据分析的思路。

地图术可以辅助设计师快速进入设计问题的思考，并通过艺术化的手段，对设计问题的

---

valuable insights into the evolution of landscape design principles. Therefore, a long-term perspective can be adopted to consider the future requirements of landscape architecture (Fig 6-18).

6.3.2 Lay involvement

In addition to the research conducted by design professionals for a project, it is necessary to incorporate the perspectives of lay individuals in order to obtain a comprehensive and holistic understanding of the design challenges. Many design projects are executed without involving the public from the initial stages of conceptualization to the final implementation. While the designer's initial intentions are often noble, with the aim of providing a pleasurable experience for the public, our own cognitive limitations restrict us from fully understanding the needs of others. Therefore, it is essential to seek the perspectives of invested non-professionals after the design process, in order to duly consider their requirements.

Questionnaires, interviews, and participatory design activities serve as valuable tools for obtaining public opinion.

Questionnaires can be administered at various stages of the project, enabling regular evaluations to be conducted. Research interviews and participatory design activities primarily take place during the pre-design phase. In-depth interviews necessitate a significant amount of time and should be conducted prior to the complete development of design ideas. During the pre-design phase, a comprehensive range of multi-perspective questions is indispensable, while the post-design phase calls for explicit and decisive design choices to guide the project towards a successful conclusion.

6.4 Information collation

Mapping refers to a method of organizing information, which comprises a systematic approach to organizing information, analyzing the content of design research, and presenting the resultant findings.[112] The process of mapping comprises a series of multifaceted procedures, including but not limited to reviewing, examining, classifying, and summarizing preliminary data (Fig 6-19). This method of cartographic representation aims to effectively convey specific data derived from maps and

图 6-18　Landscape design in timeline analysis
　　　　基于时间轴分析的景观项目

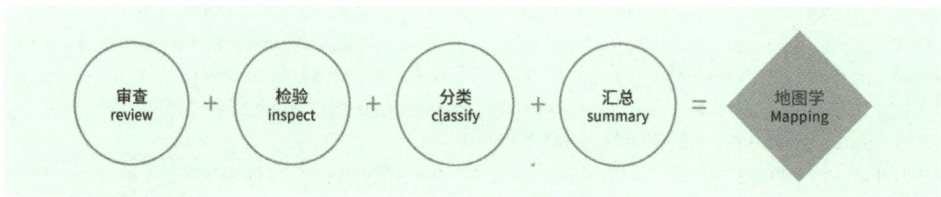

图 6-19　Mapping process
　　　　地图术分析过程

具体解决方法进行导向式的科学分析，为下阶段设计策略提供真实的研究基础（图 6-21）。

小结

通过了解景观设计调查和分析方法，掌握不同调研方式的优缺点，并且理解调研方法的使用目的，从而在设计实践的过程中选择合适的调研方法进行问题分析。调研方法和分析系统是解决设计问题的具体工具，学生需要在设计过程中对场地进行综合的观察和判断。

research through the utilization of impactful infographics. Design analysis integrates both rational and emotional processes, eschewing excessively rigid design decisions while duly considering human and social factors. Overreliance on data-driven design decisions, without the provision of reasonable feedback on public needs, may contribute to unfavorable results. The visual expression of research empowers designers to explore a myriad of creative techniques in order to identify multiple problem-solving approaches. For instance, particular emphasis is placed on the historical heritage of a given site, and the temporal evolution of urban renewal is explained through a timeline of historical scenes. Building upon this foundation, the interrelationship between scenes and human activities is assessed from the perspective of employing cartography (mapping) as a means of creating elements, including the layout of building facades, pavement texture, and recreational amenities (Fig 6-20). In addition, the design research literature frequently conveys the challenges encountered and the conventional methodologies employed by designers in their exploration of case studies. Mapping serves as a valuable tool in enabling designers to swiftly understand design problems. By employing artistic means, it facilitates a scientifically oriented analysis of specific design solutions, thereby providing a robust research foundation for subsequent phases of design strategies (Fig 6-21).

Summary

Students will be introduced to various research methods, research evaluation criteria, and research tools, thereby acquainting them with the advantages and disadvantages associated with each. Moreover, they will gain an understanding of the purpose behind employing research methods, thereby enabling them to judiciously select the appropriate research methodologies for problem analysis in their design practice. Research methods and evaluation systems represent specific instruments accorded to students to address design problems, necessitating comprehensive assessments based on site observations during the analysis phase.

图 6-20　**Mapping of Suzhou River**
苏州河地图分析

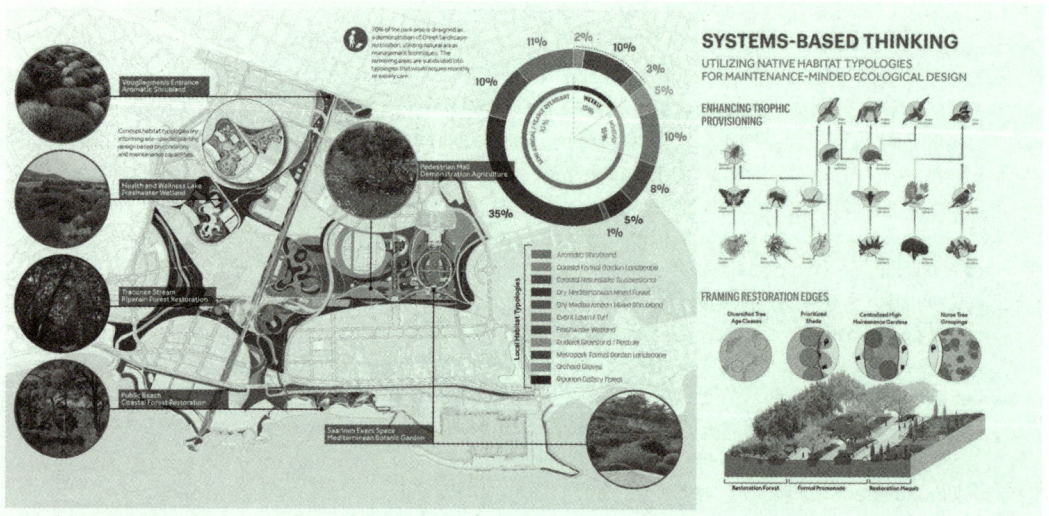

图 6-21　**Landscape mapping**
景观场地地图分析

课后作业
After-class exercises

———

讨论题
Discussion

针对设计调研的不同方法，请结合调研的经验，分析其中一种方法的优势和劣势。

Do students analyze the advantages and disadvantages of different methods of designing research in relation to their research experiences?

设计实践
Design Practice

结合数字地图技术与场地观察，选择生活中熟悉的区域，使用地图术分析景观场地与城市、人、社区的关系。

Combine digital mapping techniques and field observation, select familiar scenes from life, and use pictorial (mapping) methods to correlate landscape locations with cities, people, and communities.

推荐读物
Reference reading

1. 何志森：《Mapping 工作坊：重新解读城市更新与日常生活的关系》，《景观设计学》2017 年第 5 期。
2. 何志森、杨薇芬：《大地之上：基于人的尺度的图绘》，《国际城市规划》2019 年第 6 期。
3. 孟彤：《景观元素设计理论与方法》，2012 年。
4. 王志芳：《景观设计研究方法》，2022 年。
5. Kopec, D. *Environmental Psychology for Design*. Fairchild Books, 2018.
6. Stahlschmidt, P., Swaffield, S. and et al. *Landscape Analysis: Investigating the potentials of space and place*. Routledge, 2017.
7. Steg, L. and De Groot, I. M. *Environmental Psychology: An Introduction*. 2018.
8. Weller, R., Drozdz, Z. and Kjaersgaard, S. P. "Hotspot cities: Identifying peri-urban conflict zones" *Journal of Landscape Architecture*, 2019(14): 8-19.

第七章
景观基础设施
与规划

Chapter 7
Landscape infrastructure planning

学习要点
Study Points

了解景观基础设施的分类，并掌握其设计规范和标准

Acquire in-depth knowledge of the classification of landscape infrastructure and master its design specifications and standards

学习目的
Teaching Objectives

通过对生态基础设施、水资源管理基础设施、文化服务基础设施、康乐运动基础设施的分类介绍，使学生了解在城市规划尺度、空间尺度、细部尺度下景观基础设施的规划原理，并掌握基础设计规划方法。

This course aims to introduce students to the classification of ecological infrastructure, water management infrastructure, cultural service infrastructure, and recreational and sports infrastructure. Students will gain insight into the planning principles of landscape design within infrastructure design at various scales, including urban planning, spatial, and detailed scales. Furthermore, students will develop proficiency in basic design planning and method design.

景观基础设施是城市生活空间的重要组成部分。景观设计已经从传统的"围合式庭院"融入城市公共空间的发展。本章节以四个类型的基础服务设施进行设计问题观察，通过案例的分析诠释景观基础设施的设计策略，同时思考景观基础设施如何应对城市发展所面临的环境挑战。

Landscape infrastructure holds significant importance in urban living spaces. Landscape design has evolved from the conventional "enclosed courtyard" concept to the development of urban public spaces. In this chapter, we explore four types of infrastructure services and elucidate the design strategies of landscape infrastructure through case studies. Moreover, we investigate how landscape infrastructure can address the environmental challenges posed by urban development.

一、生态基础设施

什么是"生态基础设施"？生态基础设施是保持、改善和增加生态系统服务必备的一系列条件和组合，类型包括城市绿地、湿地、农田、生物滞留池、绿色屋顶等自然和半自然系统，是城市可持续发展的重要保障和支撑。[113]

生态基础设施按服务功能可以分为四大类型：生态调节服务设施、文化娱乐服务设施、景观支持服务设施、供应服务设施。

1. 生态调节服务设施

生态调节服务设施是主要针对区域气候调节、空气调节、水循环调节、土壤净化、生物控制和噪声防护所建立的景观基础设施，[114]如自然森林公园、湿地公园、城市绿化带、小型水体和溪流、雨水花园、屋顶花园、可渗透铺装等。

（1）自然生态系统的服务设施设计策略

巴哈马群岛莱昂李维原生植物保护区

地点：巴哈马｜面积：12.14公顷（30英亩）

时间：2021

设计公司：Raymond Jungles

巴哈马群岛是世界上生态系统面临最大风险的地带，岛上以本地植物为主，人居生活带来的污染威胁着当地景观环境的质量。项目通过将非法垃圾场改造成具有旅游和教育意义的公园场所，利用生态修复的方法把农业蓄水池转换为淡水湿地和能观赏体验的花园步道，不仅提升了生态环境的质量，也为宣传环境保护意识提供了直接的体验和教育场所。项目坐落在巴哈马国家公园保护区，园区内有丰富的本地植物，特别是传统药材。为了保护本地物种和生态栖息地免受森林砍伐导致的环境侵害，

7.1 Ecological infrastructure

"Ecological infrastructure" encompasses the various conditions and combinations necessary for the maintenance, enhancement, and augmentation of ecosystem services. These services revolve around natural and semi-natural systems (e.g. urban green spaces, wetlands, farmlands, bioretention ponds, and green roofs). Ecological infrastructure plays a pivotal role in effectively safeguarding and promoting sustainable urban development.[113] It can be categorized into four primary types based on their functions: ecological regulation facilities, cultural and recreational facilities, support facilities, and supply facilities.

7.1.1 Ecological regulation facilities

Ecological regulating facilities pertain to infrastructure dedicated to regional climate regulation, air quality management, water cycle regulation, soil purification, biological control, and noise mitigation.[114] These services are largely manifested in the landscape design of natural forest parks, wetland parks, urban green belts, small water bodies and streams, rain gardens, roof gardens, permeable paving, and various other landscape design elements.

Design strategies for service facilities in natural ecosystems

Leon Levy, Bahamas, maintains a native plant reserve

Location: Bahamas | Area: 12.14 hectares (30 acres)

Date: 2021

Designed by Raymond Jungles

The Bahamas is home to one of the world's most vulnerable ecosystems, primarily consisting of native plants. Pollution stemming from human communities poses a significant threat to the local environment. This project has successfully transformed an illegal dump into a parkland site, taking on both tourism and educational significance. Through the application of ecological restoration techniques, an agricultural impoundment has been repurposed into a freshwater wetland and a garden trail. The project not only enhances the quality of the ecological environment but also offers a visual and experiential educational venue to raise environmental awareness.

Situated within the Bahamas National Park Reserve, known for its rich array of native plants, including traditional medicinal herbs, the project's primary focus lies in the

项目主要采取了三大设计策略。第一，对保护建筑和具有旅游功能的建筑空间进行场地规划，确定合理便捷的交通路线和建筑功能组织，为园区的运行提供基础保障（图7-01）。第二，体验路线与生态保护场所相结合（图7-02）。在步行和骑行的路线上，参观者经过原生植物的灌木丛，能直接观察迁移后的石头和椰子树，石头和植物形成的古老肌理让参观者感受到历史的痕迹（图7-03）。第三，改造本土植物养殖中心和原有蓄水池，恢复原生植被，让杂乱无章的废弃植被得到合理的养殖和保护（图7-04）。重新得到保护的原生植被被改造成无障碍花园，既能保持多元物种协调发展，又能吸引更多的动物栖息于此。

最后，设计师还加强了项目的可持续运营理念，在项目外扩建新的旅游中心、学生和教师研发中心，为生态环境的持续研究提供稳定的观察基地（图7-05）。

* 该项目荣获2021年美国ASLA通用设计类荣誉奖

广东清远飞来峡海绵公园景观设计

地点：中国广东｜面积：0.24公顷（0.6英亩）

时间：2018

设计公司：GVL怡景国际设计集团

通过生态修复的方式进行设计更新，景观设计师需要面对时间和成本上的挑战。生态修复是通过自然生态圈的平衡，让土壤、微生物、植物进行自我养护和更新的设计方法。但修复时间较长，无法得到立竿见影的效果，这需要设计师和政府都具有坚定的信心。生态修复所面对的项目面积也较大，需要考虑如何在低成本的建造基础上达到最大的影响范围。

protection of native species and ecological habitats from the adverse effects of deforestation. It employs three key design strategies. First, it involves site planning for conservation buildings and spaces dedicated to tourism functions, carefully selecting reasonable and convenient transportation routes while organizing buildings to facilitate the park's operation (Fig 7-01). Second, it integrates walkways with ecological protection areas, allowing visitors to pass through native plant zones and observe relocated stones and coconut trees along the walking and riding routes (Fig 7-02). The ancient texture created by the stones and plants enables the visitors to connect with the historical heritage of the area (Fig 7-03). Third, the project addresses the preservation of the overgrown and abandoned vegetation environment by renovating the native plant breeding centre, restoring the original water reservoir, and revitalizing the native vegetation (Fig 7-04). This protected native vegetation has been transformed into an accessible garden, while the original vegetation continues to harmoniously support multiple species, serving as habitats for animals and insects.

Additionally, the project's designers emphasize the concept of sustainable operation by expanding the new tourism, student, and faculty research and development centre. This expansion lays the foundation for ongoing ecological research and development efforts (Fig 7-05).

* The project received a 2021 ASLA Honor Award in the Universal Design category.

Guangdong Qingyuan Feilai Gorge Sponge Park Landscape Design

Location: Guangdong, China | Area: 0.24 hectares (0.6 acres)
Duration: 2018
Designed by GVL Yijing International Design Group

Landscape architects must grapple with time and cost challenges when engaging in ecological restoration for renewal. The ecological restoration design approach represents a process aimed at nurturing and rejuvenating soil, microorganisms, and plants while maintaining the delicate balance of the natural ecosystem. However, restoration is a lengthy endeavour, and immediate

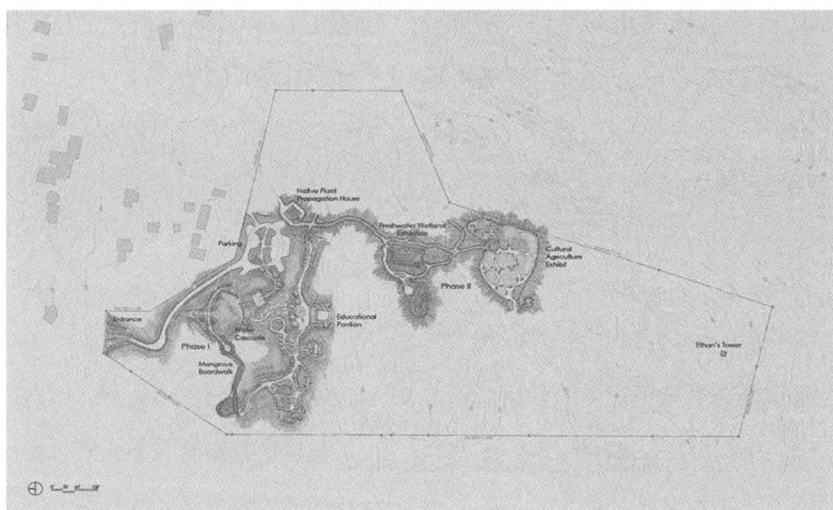

图 7-01　**Site plan**
场地平面图

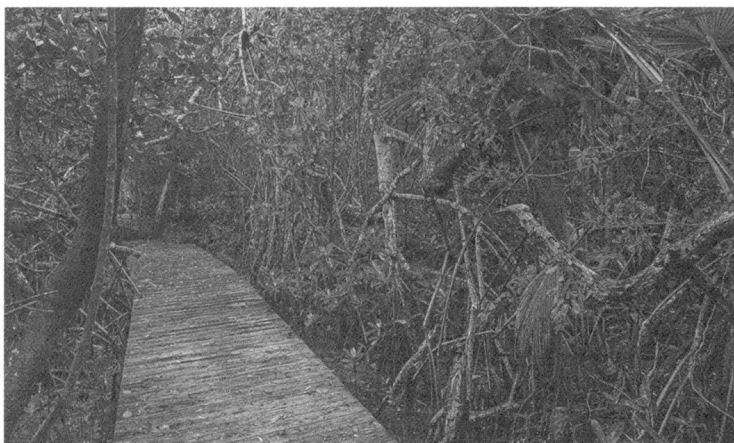

图 7-02　**Mangrove tree trestle road**
红树林木栈道

图 7-03　**Trail detail**
小径细节

图 7-04　**Native plant breeding**
本土植物繁育室

图 7-05　**Teaching and advocacy activities for children**
儿童教学和宣传活动

广东清远的飞来峡海绵公园项目就根据低成本和低影响（Low Impact Development，简称 LID[115]）的设计理念对场地进行生态修复（图 7-06）。通过情景模拟的分析方法（图 7-07），对人工湿地、雨水湿地、人工浮岛、湿地浮岛这四种雨水管理模型进行变量化分析，从而分析每种雨水管理系统的低维护建造、低维护成本、功能维度和服务寿命。场地作为污水处理场户外的绿地空间，将雨水管理（图 7-08）和污水处理（图 7-09）相结合，形成以植物物种和基质（图 7-10）为基础的净化循环系统。项目还结合了多种科学的生态测试方法，对污水处理的效率进行定量分析和测评，对选定的具有净化作用的植物进行表现能力的数据测试。

该项目的研究意义具有较强的普适性，针对切合实际的雨水管理的营造方法、成本控制、材料选择都提出了符合本地条件的方案。对于其他地区应用同样的研究方法进行设计具有较强的参考价值（图 7-11）。

* 该项目荣获 2021 年美国 ASLA 研究类荣誉奖

（2）城市公共空间

美国波士顿"翡翠项链"规划

地点：美国 ｜ 面积：约 450 公顷（1111 英亩）

时间：1878 至今

设计师：弗雷德里克·劳·奥姆斯特德

美国波士顿"翡翠项链"项目[116]，是近现代城市绿道公园的创新，同时也是生态环境修复和可持续景观发展的成功典范。它给整个城市绿色生态环境带来了巨大的提升，并持久地影响着波士顿城市公共空间的发展（图 7-12）。波士顿"翡翠项链"的规划项目从 19 世纪开

results are not feasible. Consequently, it necessitates the exchange of information between designers and government bodies. Moreover, ecological restoration often encompasses extensive project areas, prompting questions regarding how to achieve maximum effect at a minimal cost. In the case of this project in Qingyuan, Guangdong, the ecological restoration strategy aligns with the principles of "low cost" and "low impact development" (LID) [115] (Fig 7-06). The site's design combines outdoor green spaces with wastewater treatment (Fig 7-07), incorporating rainwater management (Fig 7-08) and wastewater treatment (Fig 7-09) to create a purification system using a combination of plant species and substrates (Fig 7-10). The project also employs a variety of scientific ecological testing methods to quantitatively assess the efficiency of wastewater treatment and to collect data on the purifying capabilities of selected plant species. Furthermore, the project conducts a variable analysis of four stormwater management models (artificial wetlands, stormwater wetlands, artificial floating islands, and wetland floating islands) using scenario-based analysis methods.

This approach leads to cost-effective construction, low maintenance expenses, and an efficient stormwater management system. The research findings from this project offer valuable insights that can be widely applied, providing practical methods for stormwater management, cost control, and material selection tailored to local conditions. These insights can serve as a valuable reference for other regions looking to employ a similar research methodology for design analysis (Fig 7-11).

* The project was awarded a 2021 ASLA Honor Award in the Research category.

Design strategies for urban public spaces
Boston Emerald Necklace Project, USA

Location: USA ｜ Area: approx. 450 hectares (1,111 acres)

Duration: 1878-present

Designed by Frederick Law Olmsted

The Emerald Necklace project[116] in Boston, USA, stands as a pioneering example of modern urban greenway parks. It serves as a highly successful model of ecological restoration

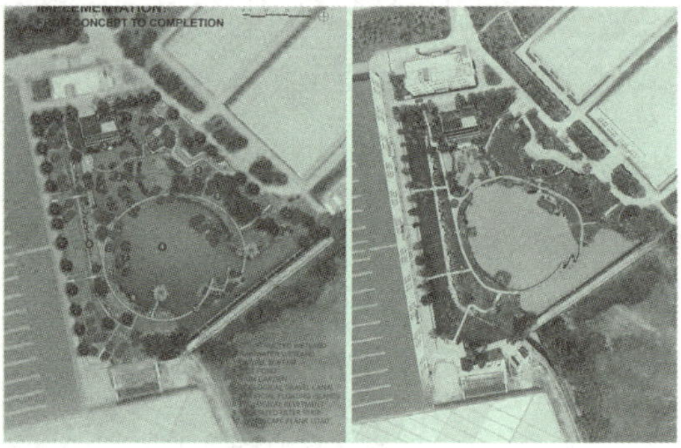

图 7-06 **From concept to implementation**
从概念到落实

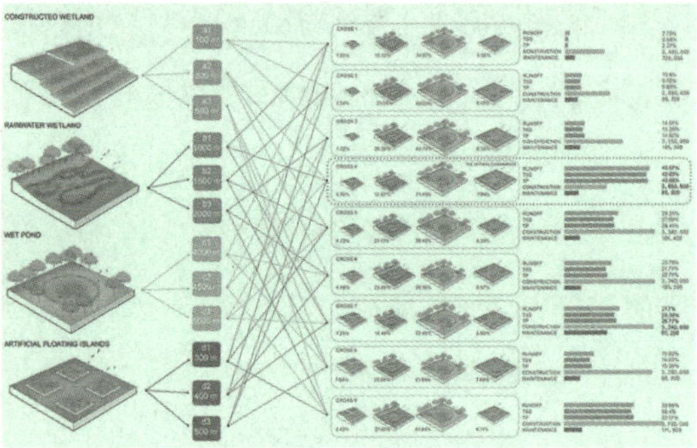

图 7-07 **Scenario simulation analysis**
情景模拟分析

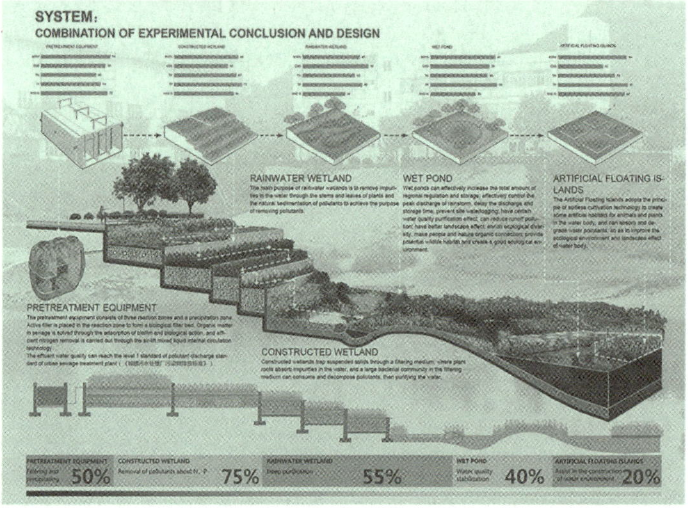

图 7-08 **Rainwater management: the experimental conclusion is combined with the design**
雨水管理：实验结论与设计相结合

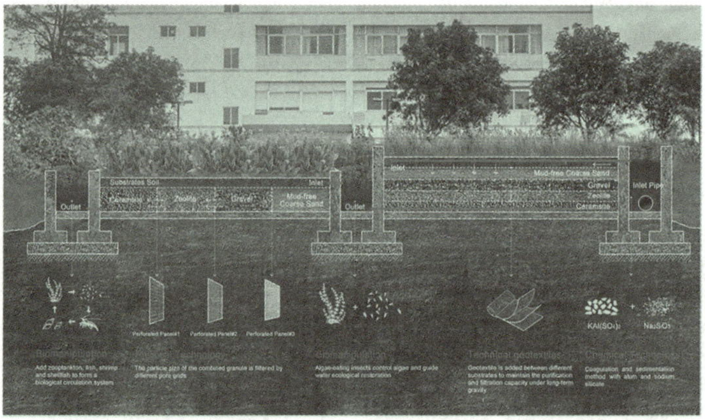

图 7-09 **Wastewater treatment**
污水处理

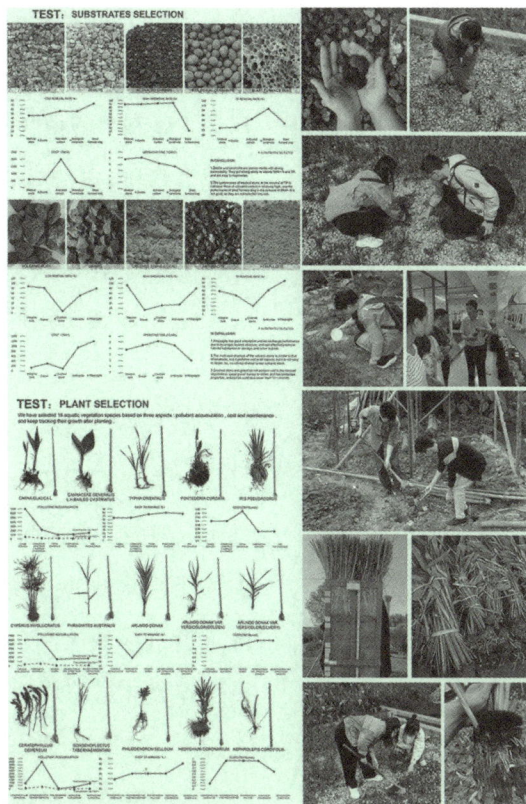

Experiment: substrate and plant selection

图 7-10   试验：基质和植物的挑选

Completion photo

图 7-11   完成后的照片

始，已历经百年考验。它的规划理念，以及面对城市变化的需求仍然能可持续地进行回应和发展，这对景观设计师来说都是活生生的教科书。

19世纪的波士顿经历了工业污染和城市人口的激增，恶劣的城市环境亟须通过建设公共花园来改善（图7-13）。奥姆斯特德认为，城市居民需要经常接触自然，不仅是他们的生理需求，更是他们的精神需求。他将公园的常规形态打破，通过约16千米的绿廊把9个公园和居民日常生活连接在一起。这个串联的连续绿道成为世界上最早的城市绿道之一，帮助城市改善了环境污染问题。参照这一点，现代波士顿的公共空间开发仍与历史公园、商业街道、火车站等各个城市点位不断地连接和延伸。

城市居民可以在车行道路中间进行运动、漫步、带儿童玩耍，并且能无打断地走到各个城市公园的中心。不同的绿廊板块有着不同的活动主题和生态植物群落，为居民的生活和享受提供便利（图7-14、7-15、7-16）。

上海徐汇跑道公园

地点：中国上海 | 面积：8.22公顷（20.3英亩）

时间：2018

设计公司：Sasaki

上海徐汇跑道公园（图7-17）的设计区别于一般城市公共绿地的设计，在城市空间中加入了雨水收集系统、生态植物设计，以及湿地营造模块（图7-18）。这也是上海市第一个建设在道路周边的雨水花园系统[117]（图7-19）。绿地下凹陷的雨水花园地形，可以对收集区的道路径流悬浮物和污染物进行生态净

and sustainable landscape development, greatly enhancing the overall urban ecological environment and contributing to the ongoing growth of Boston's urban public spaces (Fig 7-12). The Boston Emerald Necklace planning project, which has been under examination since its inception in the nineteenth century, embodies a planning philosophy and a capacity to adapt and develop sustainably in response to evolving urban needs. It can be considered a living textbook for landscape architects.

During the nineteenth century, Boston faced the challenges of industrial pollution and a rapid urban population influx. The dire urban environment necessitated the creation of public gardens (Fig 7-13). Designer Frederick Olmsted recognized that city dwellers require regular access to nature for their physical needs and their spiritual needs. In response, he departed from conventional park design by connecting nine parks through a green corridor stretching nearly 16 kilometres, intimately linking these parks with the residents' daily lives. This continuous greenway earned recognition as one of the world's original urban greenways, significantly mitigating the city's environmental pollution problems. The modern development of Boston's public spaces remains intertwined with a variety of urban locales (e.g. the city's other historic parks, commercial streets, as well as train stations).

Within this extensive network, city residents enjoy the ability to engage in activities such as exercise, strolls, and play with their children right in the middle of the carriageway. They can traverse to the heart of various city parks without interruption. Each segment of the green corridor features distinct activity themes and ecological plant communities, further enhancing the convenience and enjoyment of the residents (See Figure 7-14, 7-15).

Xuhui Runway Park

Location: Shanghai, China | Area: 8.22 hectares (20.3 acres)
Duration: 2018
Designed by Sasaki

The Xuhui Runway Park project (Fig 7-17) in Shanghai distinguishes itself from traditional urban public green space designs by incorporating rainwater harvesting,

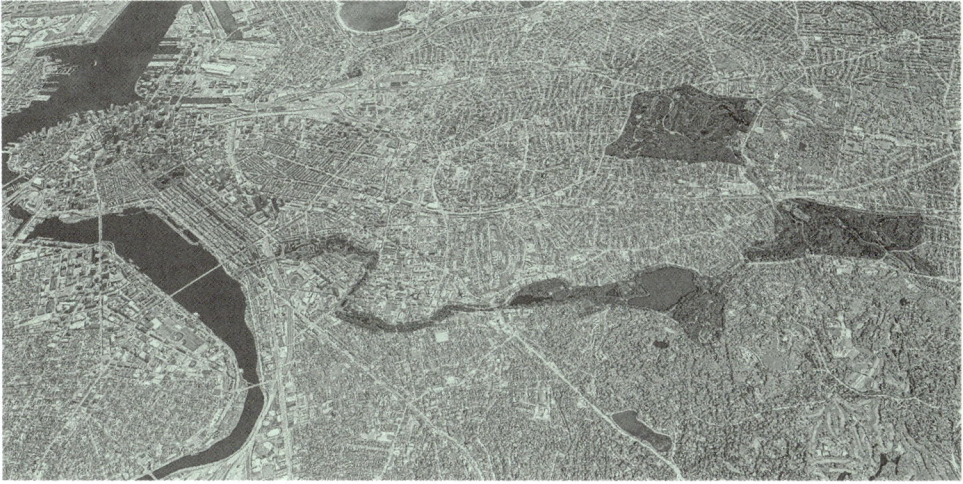

图 7-12   **Site plan**
          场地平面图

图 7-13   **A crowded Boston in 1879**
          1879 年拥挤的波士顿

图 7-14   **Urban parkway**
          城市公园绿道

图 7-15   **Water feature**
          景观水景

图 7-16   **Running track**
          运动路径

化，避免直接排入黄浦江而造成水资源污染。

同时，在植物设计上，选取长三角本地植物，形成适合陆生、水生的植物群落。通过丰富的植物群落营造，构成适合动物和昆虫栖息的城市绿地环境（图7-20）。观鸟园、果林、湿地景观场景的设计，让日常街道景观展示出自然季节变化的奇妙。

* 该项目荣获2021年美国ASLA城市设计类荣誉奖

（3）城市绿化带和雨水花园设计策略

猎人角南区滨水公园（二期）

地点：美国纽约 | 面积：4.45公顷（11英亩）

时间：2008—2012

设计公司：Arup SWA / Balsley / Weiss / Manfredi

该项目位于纽约长岛地区，将受污染的棕地环境[118]转化为可持续的示范性滨水空间（图7-21）。设计方案将景观设计与基础设施进行

创新性结合，公园融入社区需要的各种休闲和文化功能，并且具有科学的协助团队应对突发性水位上涨。项目采用了"柔软"的绿地适应性设计，首先对堤道进行绿色处理，设计可被淹盖的绿地系统，从而减少水位上涨对堤岸的破坏。其次，将原有的垃圾填埋半岛区域重新建造为沼泽栖息地，从而形成纽约最新的岛屿（图7-22）。步行栈道、生态湿地、野餐区域和观景场所拓展了人们对于滨水空间的认识（图7-23、7-24、7-25）。

该项目是可持续设计方法的重要体现，其规划策略和景观空间的细节处理都体现了滨水空间在城市堤岸保护、本地植物生态保护、社区保护方面的综合作用。[119]

* 该项目荣获2019年美国ASLA通用设计类荣誉奖

ecological plant design concepts, and wetland creation modules within the urban landscape. It represents the first rain garden system established alongside a road in Shanghai (Fig 7-18). Leveraging the recessed rain garden[117] topography within the green space, runoff containing suspended particles and pollutants from the road is directed towards a collection area for ecological purification (Fig 7-19). This approach helps prevent direct discharge into the Huangpu River.

Moreover, the design carefully selects indigenous plant species from the Yangtze River Delta, creating a plant community well-suited for both terrestrial and aquatic environments. This contributes to an urban green space that provides a hospitable habitat for animals and insects through the establishment of diverse plant communities (Fig 7-20). In addition, the design incorporates elements such as bird-watching gardens, fruit forests, and wetland landscapes, allowing everyday streetscapes to showcase the marvels of seasonal changes in the natural environment.

Design strategies for urban green belts and rain gardens
Hunter's Point South Waterfront Park (Phase II)

Location: New York, USA | Area: 4.45 hectares (11 acres)
Duration: 2008–2012
Designed by Arup SWA/ Balsley / Weiss/Manfredi

Situated in the Long Island area of New York, this project entailed the transformation of a contaminated brownfield[118] environment into an exemplary waterfront space characterized by sustainable design (Fig 7-21). The innovative design solution involved a harmonious blend of landscape design and infrastructure, resulting in a park that incorporates diverse recreational and cultural amenities to cater to the community's needs. Notably, it also introduced a "soft" green space adaptation to address abrupt water level rises along the waterfront. The "soft" green space adaptation is initiated through a green approach to the causeway, where a floodable green space acts as a protective buffer against potential damage caused by rising water. Furthermore, the original landfill peninsula area has undergone reconstruction to create New York's newest island (Fig 7-22). This newly developed

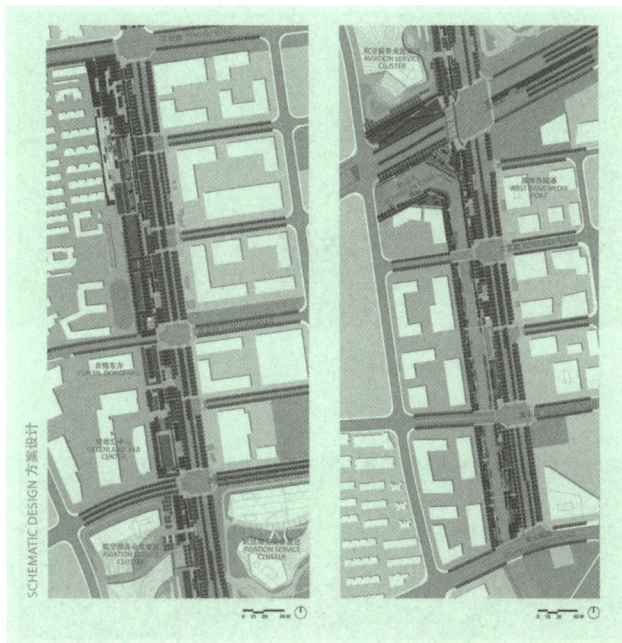

图 7-17 Site plan
场地平面图

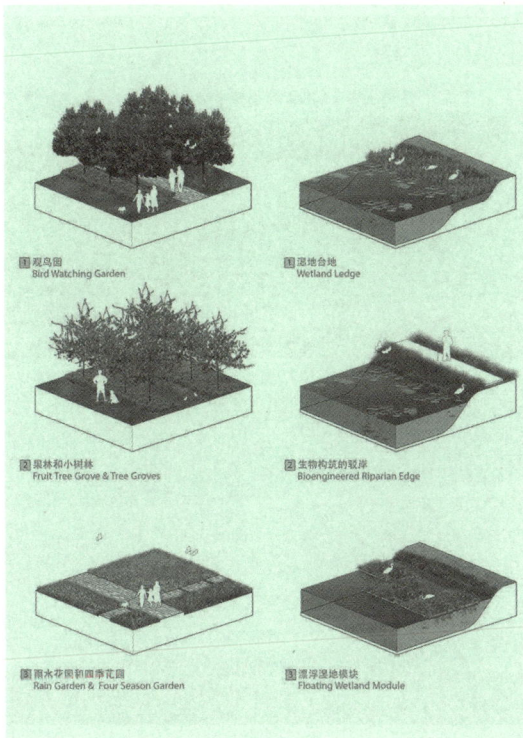

图 7-18 Design module for wetland
湿地营造模块

图 7-19 Rainwater harvesting
雨水收集

图 7-20 Birdview of the park
公园俯瞰图

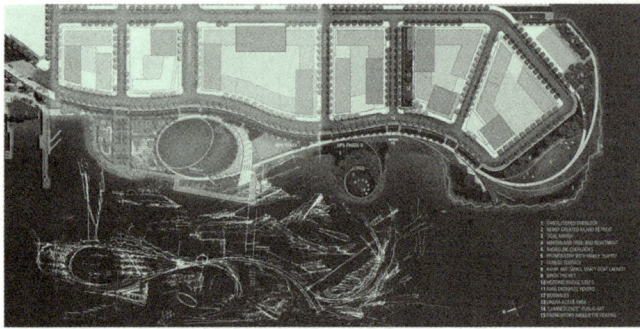

图 7-21 **Plans and design sketches**
平面图和设计草图

图 7-22 **Sustainable development map**
可持续发展示意图

图 7-23 **Wetland steps**
湿地台阶

图 7-24 **Promontory and wooden walkway**
海湾高地和木栈道

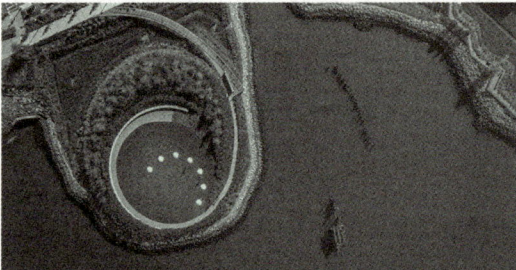

图 7-25 **Small island reserve**
小岛保护区

杜克大学雨水花园

地点：美国 | 面积：2.23 公顷（5.5 英亩）

时间：2010—2015

设计公司：Nelson Byrd Woltz Landscape Architects

该项目位于美国南部北卡罗来纳州的杜克大学校园。当地因为长期受到干旱和暴雨的极端气候影响，产生了大面积的雨水径流冲刷，并且缺少雨水收集的城市系统。如何利用暴雨带来的雨水应对干旱天气的用水紧缺，是该项目对于环境气候变化的设计思考。

设计师在杜克大学校园内建造雨水管理池，通过地形高差营造具备蓄水功能的滨水空间（图 7-26）。在蓄水池周边设置植物观察站、休闲步道、休闲平台供学生进行体育休闲活动（图 7-27）。围绕休闲步道搭配种植 40 种以上能应对极端天气的本地植物，丰富全年不同季节的视觉景观，也降低了景观设计中水资源灌溉的成本。杜克大学雨水花园的更新体现了景观基础设施连接教育、自然、社区的重要使命，也是可持续设计的重要典范。

（4）绿色屋顶设计策略

绿色屋顶对于缓解城市热岛效应有着重要的作用。[120] 面对大面积的硬质屋面，如何进行有效且可持续的绿地设计，是当代设计师需要面对的紧迫问题。对于屋顶绿化的设计，国内外的实践已经从简单的绿化种植向社交功能、生态保护、建筑设计的不同方向发展，如同时具有低能耗和生态保护功能的屋顶花园。澳大利亚墨尔本伯克利屋顶花园选择低维护成本的植物进行多样化的种植设计，并通过地形营造多个可以进行雨水收集的区域，为屋顶设置可循环雨水灌溉系统。屋顶花园使用的材料也秉

area boasts walking trails, ecological wetlands, picnic spots, and scenic viewpoints that significantly enhance the overall perception of the waterfront space (Fig 7-23, 7-24, 7-25).

This project serves as a significant embodiment of a sustainable design approach, reflecting integrated planning strategies and landscape treatment details. It not only functions as an urban waterfront emergency system but also prioritizes the ecological protection of native plant species and the preservation of the local community.[119]

* This project received a 2019 ASLA Honor Award in the Universal Design category.

Duke University Rain Garden

Location: North Carolina, USA | Area: 2.23 hectares (5.5 acres)
Duration: 2010-2015
Designed by Nelson Byrd Woltz Landscape Architects

This project is situated on the campus of Duke University in the southern state of North Carolina, USA. The local environment faces the challenges of prolonged droughts and heavy rainfall, leading to substantial stormwater runoff and a lack of urban systems for rainwater harvesting. The project's design tackled the issue of rainwater harvesting from heavy rainfall to address water shortages during dry spells.

Designers initiated the creation of a rainwater management pond on the Duke University campus by altering the topographic elevation, effectively forming a waterfront space for water storage (Fig 7-26). This innovative approach is accompanied by the establishment of plant observation stations, recreational trails, and recreational platforms, offering students opportunities for physical recreation (Fig 7-27). Additionally, more than 40 species of native plants, known for their resilience in the face of extreme weather conditions, have been strategically planted along the trail. This not only enriches the landscape perspective across different seasons but also reduces the costs associated with water irrigation.

The revitalization of Duke's rain garden exemplifies the critical role of landscape infrastructure as a bridge

承可持续的设计理念，用原木、瓦片和树枝构筑鸟类与昆虫栖息的地方。该项目虽然面积较小，但为屋顶花园可持续设计策略提供了较高的参考价值。

设计师对于自然有着无限的追求和向往。从巴比伦的空中花园[121]到现代的垂直森林建设，景观设计师和建筑师始终在为不同的建筑形态创建尽可能多的绿色空间。意大利米兰的垂直森林（图7-28）和上海的千树广场（图7-29），都展现了对建筑空间可持续性的理解，以及对如何利用有限的空间拓展绿色生态的发展，让城市的建造环境融入自然生态的思考。另外，巴黎夏蒂埃·达利克斯科学与生物多样性小学和泰国国立法政大学的屋顶花园，则是通过参与式种植的方式，让屋顶绿化成为拉近人与自然的公共空间。

墨尔本伯克利屋顶花园
地点：澳大利亚｜面积: 0.05公顷（约0.12英亩）
时间：2013
设计公司：Hassell

米兰垂直森林
地点：意大利｜面积: 0.85公顷（2.1英亩）
时间：2014
设计公司：Stefano Boeri

千树广场
地点：中国上海｜面积: 30公顷（74.13英亩）
时间：2022
设计公司：Thomas Heatherwick

巴黎夏蒂埃·达利克斯科学与生物多样性小学
地点：法国巴黎｜面积: 0.52公顷（1.28英亩）
时间：2014
设计公司：Chartier Dalix Architectes

connecting education, nature, and the community. Furthermore, it stands as a prominent demonstration of sustainable design development.

Design strategies for green roof design

Green roof design plays a critical role in mitigating the urban heat island effect.[120] Especially within the context of large hard roofs, it is essential to consider effective and sustainable design strategies. This is an urgent environmental issue that contemporary designers face. The practice of designing green roofs has evolved from simple greenery planting to encompass various dimensions, including social functions, ecological protection, and architectural design. For instance, rooftop gardens are now designed with a focus on energy efficiency and ecological protection. The Berkeley Roof Garden in Melbourne, Australia, stands as a prime example. This project features a diverse selection of low-maintenance plants and incorporates rainwater collection areas that align with the topography. Additionally, it has established a recyclable rainwater irrigation system for the roof. Sustainable materials are employed in the rooftop garden, including natural logs, shingles, and branches, to create a habitat for birds and reptiles. Despite its compact size, this project provides a valuable reference for sustainable design strategies concerning rooftop gardens.

Landscape designers and architects are actively creating green space for a variety of building types. From the Hanging Gardens of Babylon[121] to modern vertical forest building as a new format of architecture. Notable examples include the Vertical Forest in Milan, Italy (Fig 7-28), and the Thousand Tree Plaza in Shanghai (Fig 7-29), both of which demonstrate a profound understanding of sustainability within the architectural spaces. These projects explore how limited green space can be leveraged to expand the development of green ecology, ultimately integrating the urban environment into the natural ecology. In addition, rooftop gardens like those found at the Châtier-Dalix Science and Biodiversity Primary School in Paris and the Legislative Affairs University in Thailand are dedicated to transforming rooftop greenery into public spaces that bring people closer to nature through participatory planting.

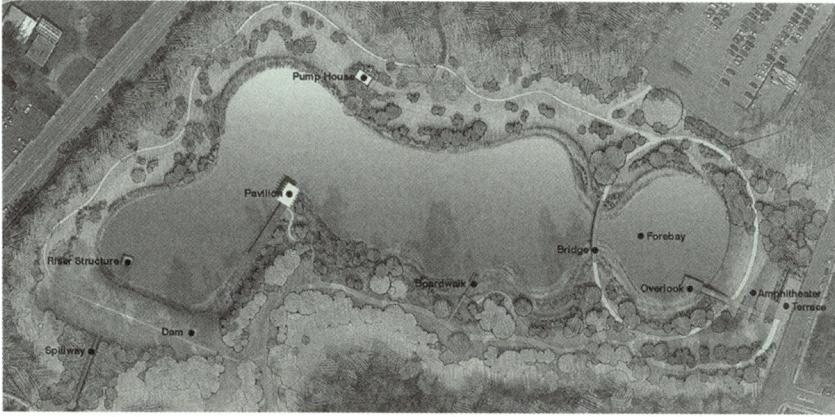

图 7-26 **Site plan**
场地平面图

图 7-27 **The main path through the site**
贯穿场地的主路

图 7-28 **Bosco Verticale in Milan, Italy**
意大利米兰的垂直森林

图 7-29 **The Thousand Trees Square in Shanghai**
上海的千树广场

泰国国立法政大学屋顶花园

地点：泰国 | 面积：2.2 公顷（5.44 英亩）

时间：2020

设计公司：LANDPROCESS (Kotchakorn Voraakhom)

（5）可渗透铺装设计策略

在景观设计中使用具有透水性的铺装材料以及雨水收集的方法都是对生态友好的设计策略。

透水材料是具有透水功能的铺设地面的材料。随着"海绵城市"[122]概念的发展，透水砖和透水混凝土材料被应用在多个城市的户外景观空间中。地表径流能迅速透过透水砖被吸收，并补充为地下水资源或者排放到雨水收集渠道中，同时，能维持城市地面的湿度和温度，避免路面积水影响行人舒适性。

另外一种"软性"透水形式为植草砖和碎石。景观设计中为了营造不同的场地气氛，会使用不同的地面材质进行铺地的设计。特别是面对停车场和消防通道这种功能特殊的地面形式，需要用更合适的设计打破大面积的硬质铺地。植草砖是通过高压砖机压成的混凝土砖，具有很强的抗压性，能承受车辆平时的通行，同时又能种植部分草地来平衡硬质铺地的面积。而碎石是景观设计中常用的造园材料，碎石之间的缝隙能快速将表面积水引流到水渠中。

2. 文化娱乐服务设施

文化娱乐和生态旅游有利于人们的身心健康，同时具有审美体验、教育、休憩等服务功能。文化娱乐服务设施包括城市森林公园、休闲绿地、长距离绿道等。以娱乐服务设施为定位的森林绿地，设计目的是达到较完整的自然

Berkeley Rooftop, Melbourne
Location: Australia | Area: 0.05 hectares (approx. 0.12 acres)
Duration: 2013
Designed by Hassell

Vertical Forest
Location: Milan, Italy | Area: 0.85 hectares (2.1 acres)
Duration: 2014
Designed by Stefano Boeri

Thousand Tree Plaza
Location: Shanghai, China | Area: 30 hectares (74.13 acres)
Duration: 2022
Designed by Thomas Heatherwick

Châtier-Dalix Science and Biodiversity Primary School
Location: Paris, France | Area: 0.52 hectares (1.28 acres)
Duration: 2014
Designed by Chartier Dalix Architectes

Rooftop Garden of Thai National University of Law and Politics
Location: Shanghai, China | Area: 2.2 hectares (5.44 acres)
Time: 2020
Design by LANDPROCESS (Kotchakorn Voraakhom)

Design strategies for permeable pavers and swales
In the construction of landscape designs, eco-friendly design strategies such as the use of permeable paving materials[122] and rainwater harvesting methods play a significant role. These sustainable approaches manifest in three design methods: permeable materials, permeable forms, and permeable rainwater swales. Permeable materials encompass substances that possess the capability to allow water to pass through when used for ground paving. The concept of the "sponge city" has gained prominence, leading to the widespread adoption of permeable bricks and permeable concrete in outdoor landscape spaces across several cities. These permeable blocks facilitate the rapid absorption of surface runoff, which can then either replenish groundwater resources or be channelled into rainwater collection systems. Moreover, they help maintain the humidity and temperature of the

风光欣赏和活动功能，设计愿景是希望体验者可以在园区进行长时间的空间体验，从而达到亲近自然和疗愈身心的作用。

如荷兰的新土地国家公园（图7-30）和德国巴特利普斯普林格森林公园（图7-31），都是在优质森林资源的基础上，引入疗养空间和森林徒步步道为主要活动区域，与生态保护区域交织在一起，如同在森林中游玩，不受人工环境干扰。

新土地国家公园

地点：荷兰｜面积：29 000公顷（约71 660.56英亩）

时间：2018

巴特利普斯普林格森林公园

地点：德国｜面积：16.8公顷（41.51英亩）

时间：2017

3. 景观支持服务设施

该景观设施旨在支持城市自然生态系统，主要体现在自然森林、湿地公园类景观空间，如野生动物栖息地、自然林地保护区、棕地修复空间等。这一类型的景观设施未必是服务于人的功能性景观设施，大部分功能设计包括绿化、水体、交通等都尽量以减少人群活动为主，尽可能给予自然生态环境最少的打扰。

4. 供应服务设施

该景观服务设施旨在提供具有绿化种植、水资源保护、动植物培育功能的景观空间，主要体现为城市绿地系统、小型水体和溪流、交通绿化带等景观区域。可以理解为规划环境下的林地种植、未被开发的原始森林绿地、自然湖泊和水源等。这类景观设施基本没有过多的人居建筑，多是未开发的自然环境，但是提供

urban ground surfaces while preventing issues related to water accumulation on pavements.

Another category is "soft" permeable forms, which include elements like grass tiles and gravel. In landscape design, a variety of ground materials are employed for paving to create diverse site atmospheres. It is noteworthy that in areas with special functions such as parking lots and fire escapes, an appropriate design is required to break up the extensive areas of hard paving. Grass tiles, created from concrete blocks compressed using high-pressure brick machines, exhibit strong resistance to pressure and are capable of withstanding vehicular traffic. They also offer the opportunity to incorporate vegetation to balance the presence of hard surfaces. Gravel, a common material in landscape design, effectively diverts surface water into drainage channels due to the gaps between the gravel particles.

7.1.2 Cultural and recreational facilities

Landscape design holds the capacity to provide not only aesthetic experiences but also educational, ornamental, and various cultural and recreational services. These

services encompass recreational and ecotourism aspects, as well as contributing to psychological, physical, and mental well-being. These functions are realized through the establishment of basic service facilities (e.g. urban forest parks, recreational green areas, and long-distance greenways). Forest green areas equipped with recreational service facilities are primarily designed to enhance the appreciation of natural scenery and outdoor activities. The overarching objective is to allow visitors to spend extended periods in these spaces, thereby reaping the benefits of connecting with nature and improving their physical and mental health.

Two outstanding examples of such high-quality forest parks are the New Land National Park in the Netherlands (Fig 7-30) and the Batlipsputger Deep Forest Park in Germany (Fig 7-31). These parks serve as healing spaces, and their forest hiking trails are central to the visitor experience. In these locations, ecological protection areas harmoniously coexist with sports and recreational activity areas, all while preserving the natural landscape without disturbance from excessive artificial structures.

通行和观察的需求驿站。

美国黄石国家公园

地点：美国怀俄明州｜面积：89.8 万公顷（约 222 万英亩）

时间：1872

二、水资源管理基础设施

可持续基础设施的建设除了以维护生态系统为目标，还可以通过水资源的管理来减轻城市对生态造成的破坏。世界各地的城市都在探索不同的水资源管理方法，主要可归纳为三个方面。

1. 水系统循环规划

在水资源循环系统中融入城市规划的理念，从宏观的层面对城市公共空间进行系统的雨水收集、雨水循环、雨水净化、雨水排放的全周期设计。在微观层面，通过城市景观空间的材料选择，在具体实践过程中加强区块绿地对于雨水管理的效率。

纽约的"The Big U"（曼哈顿下城保护计划）[123] 是丹麦先锋建筑事务所 Bjarke Ingels Group (BIG) 提出的对未来城市的畅想。在全球气候变化的条件下，纽约作为历史悠久的城市之一，应如何面对环境的危机和城市空间需求的变化？促使该事务所展开思考的是 2012 年飓风给纽约曼哈顿造成了 190 亿美元的经济损失。研究预计，到 2100 年时，海平面将上升 1.8 米，曼哈顿下城将近 20% 的街道可能会被淹没（图 7-32）。相似的情况也出现在意大利威尼斯，随着海平面逐渐上升，威尼斯每年雨季都会被海水淹没大部分的城市街道和广场（图 7-33）。"The Big U"则是通过改造景

New Land National Park
Location: Netherlands | Area: 29,000 hectares (71,660.56 acres)
Duration: 2018

Bad Lippspringe Forest Park
Location: Germany | Area: 16.8 hectares (41.51 acres)
Duration: 2017

7.1.3 Support facilities

This type of landscape facility serves to provide services within natural ecosystems (e.g. wildlife habitats, natural woodland reserves, and brownfield restoration spaces). These services are primarily offered within the landscape spaces that are part of natural forests and wetland parks. Most landscape facilities (e.g. greenery planting, water bodies, and design of traffic routes) are meticulously designed to minimize human activities and reduce any disruptions to the natural environment.

7.1.4 Supply facilities

Supply facilities are responsible for providing essential functions related to planting, freshwater resources, and landscape space for plant and animal cultivation. These facilities are primarily manifested in landscape areas (e.g. urban green space systems, small water bodies and streams, and transportation green belts). They often involve woodland planting within planned environments, the maintenance of undeveloped original forests, and the preservation of natural lakes and water sources. In general, the landscape facilities in this category do not involve excessive human settlement structures. They are typically undeveloped natural environments, made accessible for people to pass through and observe, serving as demand stations for the various elements of nature.

Yellowstone National Park, USA
Location: Wyoming, USA | Area: 898 thousand hectares (approx 2.22 million acres)
Duration: 1872

7.2 Water management infrastructure

Sustainable infrastructure can effectively mitigate ecological damage caused by cities through water management approaches. Various cities worldwide are

观基础设施的方式，为曼哈顿城提出了大胆又符合旧城更新需要的设计策略。抵御洪水的方法并不是简单地加高防护堤，而是通过对滨水空间进行韧性化处理，让滨水空间成为"软性"的防护堤，对洪水具有自我吸收和缓冲的作用。同时，根据每个区块滨水空间的建筑功能，设计符合社区活动背景的文化娱乐基础设施。在没有遇到洪水的时候，景观基础设施是居民公共空间的滨水拓展场所，当遇到洪水侵害的时候，景观基础设施又可以成为吸收和消化水量的水资源管理系统（图 7-34）。

"The Big U"受到纽约政府和基金会的大力支持，多个国际事务所共同承担不同阶段的滨水空间更新。现有建成项目有皇后区猎人角滨水公园、威廉斯堡的多米诺公园（图 7-35）、布鲁克林大桥公园（图 7-36）等。"The Big U"

为城市水资源管理基础设施的设计方法提供了多元的参考。

The Big U——曼哈顿下城保护计划

地点：美国纽约｜面积：5 公顷（约 12.36 英亩）

时间：2019

设计公司：BIG

2. 城市水资源景观空间提升

城市中的水资源空间是城市重要的自然资源。临近海湾、江河、湖泊的城市具有亲近水系的地理优势，也拥有独特的景观资源。通过提升城市水资源景观空间，可以激活城市片区的旅游、文化、商业、娱乐活动。滨水空间丰富的城市文化活动，可以促进市民对城市的归属感。位于美国芝加哥城的芝加哥河原来是沼泽溪流，但随着城市工业化的进程，在过去的 30 年间，沿河商业和居住中心不断发展。

exploring different strategies for water management governance. Among them, three key methods can be summarized as follows.

7.2.1 Water system cycle planning

This approach involves the integration of the water cycle system into the concept of urban planning. It encompasses the holistic design of systematic rainwater collection, circulation, purification, and discharge within the framework of urban public space. At the micro level, materials for urban landscape spaces are carefully selected to enhance the efficiency of rainwater management within district green areas.

The Big U (New York City Waterfront Project)[123] was conceived by the pioneering Danish architectural firm Bjarke Ingels Group（BIG）, innovatively addressing New York City's response to the environmental crisis and evolving urban spaces in the context of global climate change. In 2012, a hurricane resulted in $19 billion in economic damage to New York City. Research shows that by 2100, a 1.8-meter rise in sea levels will inundate nearly

20 percent of the streets in lower Manhattan (Fig 7-32). A similar scenario has been experienced in Venice, Italy, where rising sea levels flood most of the city's streets and squares annually during the rainy season (Fig 7-33).

The Big U proposal represents a bold and practical strategy for regenerating Manhattan by transforming landscape infrastructure. Rather than merely raising levees' heights, the design strategy aims to make the waterfront spaces resilient. These "soft" levees are designed to absorb and buffer flooding. Moreover, cultural and recreational infrastructure is tailored to align with the context of community activities according to the architectural function of the respective block's waterfront space. In non-flooding scenarios, the landscape infrastructure serves as an extension of residential public spaces along the waterfront. However, in the event of flooding, it functions as a water management system, capable of absorbing and assimilating the water.

The Big U project has garnered substantial support from the New York government and various foundations to

图 7-30　**Nieuw Land National Park**
新土地国家公园

图 7-31　**Forest Park in Bad Lippspringe, Germany**
德国巴特利普斯普林格森林公园

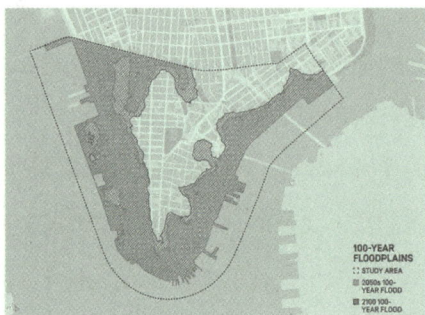

图 7-32　**100-year floodplains**
100 年后的淹没面积

图 7-33　**Venice streets flooded**
威尼斯街道被淹没

图 7-34　**The Big U design scheme**
"The Big U" 曼哈顿下城保护计划设计方案

图 7-35　**Domino Park in Williamsburg**
威廉斯堡的多米诺公园

图 7-36　**Brooklyn Bridge Park**
布鲁克林大桥公园

面对城市产业形态的转变，城市公共空间的改善和休闲文化活动的提升成为芝加哥滨水空间急迫的更新需求。芝加哥滨水空间的更新除了在建筑空间形态上创造了丰富的亲水和观赏平台，让市民感受自然的风光，还举办以水为主题的文化活动，将滨水资源转化为芝加哥的文化旅游资源。沿着芝加哥河，有浮游湿地公园供儿童进行植物观察，有皮划艇可进行户外活动，更有乘船旅游与建筑文化相结合打造的堪称"芝加哥名片"的经典建筑旅游路线（图7-37）。在春天，芝加哥河还会举办一年一度的圣帕特里克节。当天，芝加哥河会被染成绿色，以庆祝该节日。此项活动是芝加哥长达数十年的传统，是芝加哥市民聚集在滨水场所的重要节日。

通过滨水空间连接自然、教育、文化、风俗、商业和娱乐活动，已经成为世界上各个国家所关注的提升城市空间品质的重要手段。韩国首尔的清溪川修复景观项目（图7-38），以及中国上海的宝山滨水空间改造等都是滨水空间提升居民城市幸福感的具体实践。

芝加哥滨水空间

地点：美国芝加哥｜面积：1.8公顷（约4.45英亩）

时间：2016

设计公司：BIG

清溪川修复

地点：韩国首尔｜面积：9.1公顷（约22.49英亩）

时间：2003—2005

设计师：Mikyoung Kim

undertake different phases of waterfront renewal in collaboration with several international firms. Completed projects include Hunter's Point Waterfront Park in Queens, Domino Park in Williamsburg (Fig 7-35), and Brooklyn Bridge Park (Fig 7-36), each offering valuable insights into infrastructure design approaches for urban water management.

The Big U—Lower Manhattan Preservation Project

Location: New York, USA | Area: 5 hectares (12.36 acres)
Duration: 2019
Designed by BIG

7.2.2 Landscape space enhancement of urban water resources

Urban water resources have been recognized as vital natural resources of cities. Cities situated near bays, rivers, and lakes possess a unique geographical advantage due to their proximity to water systems, which serve as distinctive landscape features. The landscape of urban water resources plays a pivotal role in enhancing the development of tourism, culture, business, and entertainment activities within urban areas. Simultaneously, it fosters a sense of belonging among residents by enriching the various activities taking place along waterfront spaces.

An excellent example is the transformation of the Chicago River in Chicago, USA. Originally a swampy stream, it has undergone significant changes through the industrialization of the city. Over the past three decades, commercial and residential centres along the Chicago River have flourished. In response to the city's industrial transformation, there has been a concerted effort to improve public space and leisure culture along Chicago's waterfront. The renewal of Chicago's waterfront has led to the creation of abundant waterfront areas and viewing platforms that allow people to experience the beauty of natural scenery. Moreover, the city's waterfront resources have become integral to its tourism industry, serving as venues for water-themed cultural activities. The Chicago River features a floating wetland park where children can observe plants, and kayaks can be rented for water exploration. Boat tours are thoughtfully combined with architectural elements, offering

深圳荷水文化基地暨洪湖公园水质净化厂上部景观设计

地点：深圳市罗湖区 | 面积：3.24 公顷（约 8.01 英亩）

时间：2021

设计公司：南沙原创建筑设计工作室

3. 以水资源为主题的公共空间拓展

除了大规模的景观规划，水资源景观设施还包括了亲水平台、游泳池和观景活动设施等。

美国纽约哈德逊河上的"55 号小岛"，是英国建筑师托马斯·海瑟维克以原 55 号码头为基地，运用混凝土树桩结构建造的海上微生态滨水体验花园。树桩结构构成的花园空间，可以承载 35 棵大树、65 种灌木，以及 290 多种植被。丰富的植物层次也为城市的鸟类和昆虫带来稳定的栖息地。设计概念旨在通过植物

原生态的营造方式对原有的工业化码头肌理进行重新阐释，并通过海上花园的形式激发市民对自然环境的关注。海上小岛犹如一座自然的观察站，让城市居民可以时常借助滨水空间的独特视角去理解自然环境的各种变化。

在三面环海的丹麦，以水为主题的各种休闲活动不断在城市滨水空间中开展。丹麦的哥本哈根海港浴场通过在海边建设公共游泳池，丰富拓展市民对户外活动的参与。这个公共游泳浴场不同于常规的游泳设施，而是集社交、跳水、享受阳光和海边风景功能于一体的滨水设施，也是丹麦新型滨水环境和亲水文化的代表。水资源环境设施的创新拓展在丹麦不断得到发展，奥胡斯港浴场（图 7-39）和漂浮小岛的设计也是设计师在探索滨水空间城市体验的尝试。

---

a unique architectural tourism experience that has become one of Chicago's iconic attractions (Fig 7-37). Additionally, in the spring, the Chicago River hosts the annual St Patrick's Day celebrations. As part of this tradition, the river is dyed green to commemorate the Irish holiday, an event that has been a decades-long tradition in Chicago and a cherished annual festival where residents gather along the waterfront.

Connecting nature, education, culture, traditions, commerce, and recreational activities through waterfront spaces has become a fundamental approach to enhancing the quality of urban spaces across the globe. Projects like the ChonGae Canal restoration landscape project (Fig 7-38) in Korea and the renovation of the Shanghai waterfront space exemplify the significance of waterfront areas in improving urban residents' well-being.

Chicago Waterfront
Location: Chicago, USA | Area: 1.8 hectares (4.45 acres)
Duration: 2016
Designed by BIG

ChonGae Canal Restoration Project
Location: Seoul, Korea | Area: 9.1 hectares (22.49 acres)
Duration: 2003–2005
Designed by Mikyoung Kim

Infrastructure publicization: Shenzen Loutus Water Culture Base
Location: Shenzhen, China | Area: 3.24 hectares (8.01 acres)
Duration: 2021
Designed by NODE Architecture & Urbanism

7.2.3 Public space expansion with water resources as the theme
In addition to large-scale landscape planning and construction projects, landscape facilities cantered around water resources encompass various elements such as waterfront platforms, swimming pools, and viewing facilities.

The "Island 55" on the Hudson River in New York, USA, stands as a noteworthy example of this approach. Designed by British architect Thomas Heatherwick, it has transformed the former Pier 55 into a marine micro-

哈德逊河 55 号小岛

地点：美国纽约 | 面积：0.97 公顷（2.4 英亩）

时间：2021

设计公司：Heatherwick Studio

哥本哈根海港浴场

地点：丹麦哥本哈根 | 面积：0.25 公顷（约 0.62 英亩）

时间：2003

设计公司：BIG + JDS

丹麦奥胡斯港浴场

地点：丹麦奥胡斯 | 面积：0.26 公顷（约 0.64 英亩）

时间：2018

设计公司：BIG

丹麦漂浮小岛

地点：丹麦哥本哈根 | 面积：0.25 公顷（约

0.64 英亩）

时间：2003

设计公司：BIG + JDS

三、文化服务基础设施

文化服务基础设施包括体育和文化用地设施。其设计的标准主要源自休闲观赏、文化娱乐活动及相关基础设施的功能和要求。设施的规划设计与当地的地形环境、建筑功能布局、植物设计等有着密切的关系。以下对奥林匹克公园、主题乐园、社区儿童公园三大文化服务设施进行介绍。

1. 奥林匹克公园的设计规划

奥林匹克公园的规划设计与一般的森林公园有着不一样的意义。奥林匹克公园秉承国际使命，绿色奥运是每个举办国在设计服务基础

ecological waterfront garden. Utilizing concrete stump structures, the island creates a garden space capable of accommodating 35 large trees, 65 shrubs, and over 290 different types of vegetation. This diverse plant life establishes a stable habitat for urban birds and insects. The design concept reimagines the industrialized wharf by incorporating plant life, and its primary objective is to address citizens' concerns for the natural environment through the creation of a sea garden. This island in the sea serves as a nature observatory, allowing city dwellers to use the unique perspective of the waterfront space to understand changes in the natural environment.

In Denmark, a country surrounded by the sea on three sides, urban waterfront spaces continuously explore a wide variety of leisure activities based on the theme of water. The Copenhagen Harbor baths in Denmark exemplify this trend. These public swimming pools are not traditional swimming facilities but waterfront facilities for socializing, diving, sunbathing, and enjoying seaside views (Fig 7-39). Moreover, they symbolize the Danish people's embrace of a new waterfront environment and culture, emphasizing their

proximity to water resources. This innovative expansion of water environment facilities is an ongoing endeavour in Denmark, with projects like the Aarhus Harbor Bath and floating islands representing design attempts to further explore waterfront spaces as urban experiences.

Hudson River Island #55
Location: New York, USA | Size: 0.97 hectares (2.4 acres)
Duration: 2021
Designed by Heatherwick Studio

Copenhagen Harbour Bath
Location: Copenhagen, Denmark | Area: 0.25 hectares (0.62 acres)
Time: 2003
Designed by BIG + JDS

Aarhus Harbor Bath
Location: Aarhus, Denmark | Area: 0.26 hectares (0.64 acres)
Duration: 2018
Designed by BIG

图 7-37　**Chicago riverwalk**
芝加哥滨水空间

图 7-38　**Cheonggyecheon waterfront**
清溪川滨水空间

图 7-39　**Arhus Harbor Bath**
奥胡斯港浴场

图 7-40　**Queen Elizabeth Olympic Park**
伦敦奥林匹克公园

设施时的重要理念。每个国家在设计奥林匹克公园的时候，都需要呼应全球环境危机问题，以低耗能、高效率和绿色回收的方式进行设施和场馆的设计。奥林匹克公园需要展示场地规划的可持续性，并作为示范性的实践项目给予世界参考。

伦敦的奥林匹克公园（图 7-40）建立在污染较为严重的工业区。对场地进行生态修复后，融合沼泽、森林、体育场、水上运动中心等多个场所形成公园绿地。在奥运会结束以后，公园被更名为伊丽莎白奥林匹克公园，并对公众免费开放。伦敦奥林匹克公园以体验式生态基础设施为网络，在休闲绿地种植大量的英国本土植物以供观赏，并设有草坪、森林、湿地一体生态水循环系统。根据海绵城市的理念，构造绿色街道、可渗透铺地、生态水沟以达到绿色基础设施的建设。另一方面，奥林匹克公园使用了棕地修复的土地，并通过建造体育公园让周边本来破败的社区环境得到来自旅游和娱乐产业的激活，从而提升了城市土地的价值，改善了居民的社区环境。奥林匹克公园的设计体现了文化娱乐设施与城市规划、社区环境、自然生态的密切关系。通过文化娱乐设施的建设还可以增强居民亲近自然的意愿。

伦敦奥林匹克公园

地点：英国伦敦 | 面积：101 公顷（约 249.58 英亩）

时间：2012

设计公司：Hargreaves Associate

2. 主题乐园的设计规划

大型主题乐园的规划具有较强的主题娱乐性，需要在园区规划、单体设施建筑设计、植

Danish floating island
Location: Copenhagen, Denmark | Area: 0.25 hectares (0.64 acres)
Time: 2003
Designed by BIG + JDS

7.3 Cultural service infrastructure

Cultural service infrastructure includes the design of facilities catering to sports and cultural sites. The criteria for designing these facilities are mainly derived from the functional requirements necessary for recreational viewing, cultural and recreational activities, and the associated infrastructure. It is crucial that the planning and design of these facilities closely consider the local topographical environment, functional building layout, and plant design. The following is an introduction to the three major cultural service facilities through the lens of Olympic Park planning, theme park planning, and community children's park design.

7.3.1 Olympic Park planning

Olympic Parks hold a crucial international mission related to the concept of the Green Olympics. Each host country should prioritize the consideration of global environmental crisis issues when planning and designing Olympic Parks. This includes the design of facilities and venues that promote low energy consumption, high efficiency, and green recycling, making the Olympic Park a demonstration of sustainability in site planning and a reference of exemplary practical projects.

For instance, London's Olympic Park was established in a formerly polluted industrial area. Through extensive ecological restoration, the site was transformed into a park and green space integrating several sports venues, including marshes, forests, a sports stadium, and an aquatic centre. After the Olympics, the park was renamed Elizabeth Olympic Park and opened to the public free of charge. The London Olympic Park (Fig 7 40) features a significant presence of British native plants, enhancing the ornamental and recreational green spaces. It also implements an ecological water circulation system that integrates lawns, forests, and wetlands. The use of materials incorporates the concept of a "sponge city," which is evident in green street design, permeable pavement design, and eco-gutter

物景观方面进行综合考虑。

在规划设计中，场地的选择需要建立在无自然灾害和危险的地方，避免地震、高压、输电管、垃圾场等对人体产生危害。另外，需要有明确的交通动线，包括清晰的车行和人行流线，避免人员密集造成安全隐患；为老人、儿童、残障人士等提供无障碍通行路线；满足足够的停车标准量。游乐设施和建筑需要符合国家和当地的设计标准，特别是游乐场中的防护、防溺水和栏杆的安全规则。在主题乐园的植物设计上，需要避免有毒或带刺的植物。靠近水面的种植区域，需要使用常绿的树种以避免落叶对湖面的污染。具体的设计细节可以参照各地的娱乐设施设计标准。

英国康沃尔郡的植物园与常规植物园不同，是可持续设计的实践典范之一。它的价值还包括对可持续设计理念的教育和传播，在整体设计中加入了文娱活动功能，让一个观赏性的植物园变成一个动态的植物博物馆。植物园的选址让一个废弃的矿坑变成种植超过 4 500 种全球植物的园区，累计上千万游客参观。体验类（话剧、研讨会、艺术秀、园艺讨论、音乐节）、观赏类（温室、展览馆）、教育类三大游园路线提供了不同类型的生态环境教育服务（图 7-41）。从设计的诞生，到后期运营的过程，"伊甸"植物园都秉承了生态可持续的重要理念。

英国康沃尔郡"伊甸"植物园

地点：英国 | 面积：15 公顷（约 37 英亩）

时间：2001

设计公司：Grimshaw Architects

---

design. Additionally, the Olympic Park reclaims land from brownfield restoration, activating emerging tourism and entertainment industries in the surrounding community environment by constructing a sports park. This, in turn, increases the value of the urban land and optimizes the community environment for residents. The design of the Olympic Park highlights the close relationship between the construction of cultural and recreational facilities, urban planning, community environment, and natural ecology. It also serves to connect residents with nature by providing cultural and recreational amenities within a green and sustainable environment.

Queen Elizabeth Olympic Park

Location: London, UK | Size: 101 hectares (249.58 acres)
Duration: 2012
Designed by Hargreaves Associate + LDA Design

7.3.2 Theme park planning

The planning of a large theme park requires a comprehensive consideration in terms of park layout, architectural design of facilities, and landscaping.

In the park layout design, the chosen site should ideally be free from natural disaster hazards, such as locations prone to earthquakes, high-voltage power transmission lines, or existing garbage dumps. Moreover, Careful attention must be paid to establishing clear traffic movement lines within the park. Well-defined traffic and pedestrian flow routes are essential to prevent safety hazards caused by large crowds. Additionally, accessible traffic routes should be provided to accommodate the needs of the elderly, children, disabled individuals, and others. Sufficient parking facilities should also be integrated to handle the park's vehicular traffic. Amusement tourism facilities and buildings should adhere to national and local design standards, which often include safety regulations for fall protection, drowning prevention, and the use of railings. The design of plant landscapes in the theme park should steer clear of toxic plants or those with thorny branches and leaves. In areas near water bodies, evergreen plant species should be selected to prevent fallen leaves from polluting the water. Specific design details can refer to the design standards for local recreational facilities.

3. 社区儿童公园的设计规划

社区儿童公园的基础设施包括儿童游乐设施、防护栏杆、休闲遮阴场所、饮水机、应急警报器等。[124] 其中，儿童游乐设施需要遵守专业标准的设计规范，符合儿童游玩的安全性。儿童游玩空间中的地面可以使用软质的防滑防撞材料，同时避免家具和挡墙具有尖锐的部分。另外，社区儿童设施可根据区域气候环境和社区人群组合的特点进行灵活配置，目标是为邻里的儿童创造可以体验环境乐趣的安全舒适的活动场所。

使用成品儿童设施进行配置是城市公共空间常用的设计方法，设计师可以通过专业的儿童设施和家具生产厂家进行预订，景观设计师只需要设计安装的活动空地即可（图 7-42）。此外，设计师也会根据环境设计特殊的游玩设施。这类游玩设施需要设计团队对细节和品质有充分的考虑和检查，从而提供与环境融合良好的游玩设施。这里的游玩设施未必是普通的滑梯或者跷跷板，还可以是地形抽象而成的攀爬设施或者水流形成的铺装设施等。云朵乐园是中国成都麓湖生态城的儿童游乐场所，设计师运用丰富的想象力，通过对水的形态进行观察，设计了一系列有趣的互动游玩设施（图 7-43、7-44）。

四、康乐运动基础设施

康乐运动设施主要指具有体育功能的休闲运动设施，其中包括田径运动场、篮球场、羽毛球场、乒乓球场等。[125] 大部分的康乐运动基础设施都规划在城市公园、社区公园、学校校园等大型公共场所。但是随着人们健康意识

The botanical garden in Cornwall, UK, distinguishes itself from conventional botanical gardens through its exemplary sustainable design in practice. Its value extends beyond the mere display of plant collections, encompassing educational outreach in sustainable design and incorporating recreational activities into the overall design, effectively transforming it into a vibrant botanical museum.

The site of the botanical garden represents a remarkable transformation, having evolved from an abandoned mine pit into a flourishing park housing an impressive collection of over 4,500 plant species. This botanical garden has attracted tens of millions of visitors, and its tour route is organized to provide various forms of ecological and environmental education, including experiential activities such as dramas, seminars, art exhibitions, horticultural discussions, and music festivals. In addition, ornamental elements, such as greenhouses and exhibition halls, contribute to the garden's educational efforts (Fig 7-41). Throughout the design process and in the post-operation phase, a fundamental commitment to the concept of ecological sustainability is maintained.

Eden Project, Cornwall, UK
Location: England | Size: 15 hectares (37 acres) | Date: 2001
Designed by Grimshaw Architects

7.3.3 Community children's park infrastructure
The infrastructure of a community children's park encompasses a range of components designed to facilitate children's play and ensure their safety and enjoyment.[124] Key elements include amusement facilities, protective railings, shaded recreational areas, water fountains, emergency service alarms, and other features that contribute to the well-being of children engaging in play activities. Among them, children's amusement facilities must adhere to professional design standards that prioritize safety. The choice of flooring material in community spaces where children play should feature soft, anti-slip, and anti-collision characteristics, minimizing the risk of injuries due to falls. Sharp angles in children's furniture and retaining walls should be avoided to prevent potential harm to children while playing. In addition, the configuration of children's facilities in the community should be flexible and adaptable, taking into account regional climate characteristics and

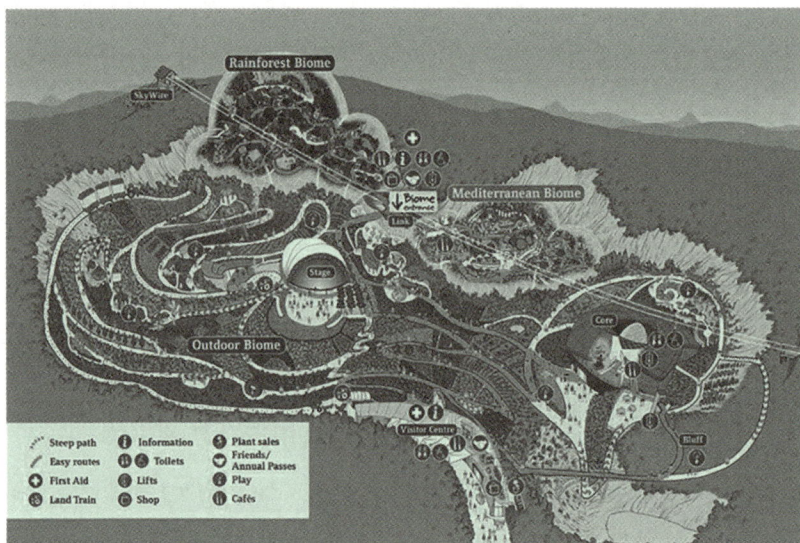

图 7-41   **Tour map of Eden Botanical Garden**
"伊甸"植物园游园图

图 7-42   **Cassiobury Park, UK**
英国卡西伯里公园

的提升，康乐运动设施也被植入城市公共空间的其他场所。

第一，在景观铺装方面，康乐运动设施通常会设计具有高弹性的软质地面，避免运动的人群因为摔倒而受伤。另外，不同的运动配置也需要不同的设计铺装材质。例如，慢跑步道和骑行步道需要设定防滑高弹性的软质地面来为运动者提供安全保护（如图7-45）。第二，小型竞技活动设施，如篮球场、体育场、羽毛球场、乒乓球场、滑板场等，设计标准必须符合体育竞技类的安全规则。在此基础上，设计师可以根据场地的特点，进行设计形式和范围的创新。例如，巴黎的彩虹篮球场在设计尺寸上满足了常规要求，但是在色彩的形式上做了大胆的创新，让运动体验与城市时尚的特点直接关联。澳大利亚的线性公园将高架桥下的灰色空间改造为公共运动场所，其中包括比11个足球场还大的开放运动空间，以及17千米长的人行道和自行车道（如图7-46、7-47）。第三，服务功能设施配置。越来越多的人选择骑行和跑步作为日常运动。运动前后所需要的服务设施配置不足，会影响运动者的体验舒适度。骑行车道和跑步路线可设置公共驿站，包括遮雨、饮水、充电、更衣、储存、卫生间功能等（图7-48、7-49）。

巴黎彩虹篮球场

地点：法国｜面积：15公顷（37.07英亩）

时间：2021

设计公司：Ill-Studio

丹德农铁道线性公园

地点：澳大利亚墨尔本｜面积：22.5公顷（55.6英亩）

the composition of the community. The goal is to create places where children can enjoy fun, safe, and comfortable activities. Design approaches for community children's parks can vary. One common method is to use finished set pieces, where designers purchase pre-built products, and landscape architects focus on designing the open spaces for their installation (Fig 7-42). Alternatively, designers can develop special amusement facilities that integrate seamlessly with the environment, taking into account details and the overall park design. These amusement facilities may not necessarily be a slide or seesaw; they can also be climbing facilities inspired by the terrain or paving facilities created by observing the water flows. Cloudland is a prime example of an imaginative children's playground in Luhu Ecological Park in Chengdu, China. Its amusement facilities revolve around a water theme, featuring interactive and engaging amusement facilities that stimulate children's creativity and playfulness (Fig 7-43).

7.4 Recreational sports infrastructure

In general, recreational sports facilities encompass a wide range of amenities, including athletic fields, basketball courts, badminton courts, table tennis areas[125], and other spaces designed for sports and physical activity. These facilities are typically located in large public places (e.g. city parks, community parks, school campuses). However, as urban dwellers become increasingly health-conscious, recreational sports facilities are being integrated into a wide variety of urban public spaces.

Recreational sports facilities are designed a focus on safety and user experience, usually using high elasticity of soft ground to minimize the risk of injuries resulting from falls. Different sports require specific flooring materials. For instance, jogging tracks and cycling trails should be equipped with non-slip, highly resilient soft surfaces to enhance safety (Fig 7-45).

Basketball courts, stadiums, badminton, table tennis, and skateboards are all competitive activity facilities and must be designed following the safety designs of each respective sports category. However, designers have the freedom to innovate within safety guidelines to create unique and visually appealing recreational sports facilities.

图 7-43　**Cloud Paradise 1**
云朵乐园 1

图 7-44　**Cloud Paradise 2**
云朵乐园 2

图 7-45　**Walking and cycling lanes**
步行和骑行车道

图 7-46 **Site plan of Dandenong Railway Linear Park**
丹德农铁道线性公园平面图

图 7-47 **Dandenong Railway Linear Park**
丹德农铁道线性公园

图 7-48 **Nike + Run Hub**
上海耐克跑者驿站

图 7-49 **Wangjiang station**
上海望江驿

时间：2018

设计公司：ASPECT Studios

望江驿

地点：上海 | 面积：0.02 公顷（0.06 英亩）

时间：2017

设计公司：致正建筑工作室

小结

　　本章节基于可持续设计的视角，结合实际案例，对生态基础设施、水资源管理基础设施、文化服务基础设施、康乐运动基础设施进行了分析和介绍。通过丰富的中外案例阐述，帮助学生了解景观基础设施的基本形态，以及所面对的多变城市空间和复杂场地关系。不同设计作品的创意提供了基础设施领域的外延扩展，从而为后续的设计实践课程积累认知基础。

For instance, the rainbow basketball court in Paris adheres to size requirements but introduces bold colour schemes, aligning the sports experience with urban fashion trends. Australia's Linear Park is a public space for sports, created by transforming underutilized spaces, like the area under the viaduct, into public sports zones. These spaces offer not only open sports areas with more than 11 soccer fields in size but also a stunning 17 kilometres of pedestrian and bicycle paths (Fig 7-46, 7-47).

With a growing number of people embracing cycling and running as part of their daily exercises, well-designed recreational sports facilities can be a game changer affecting their exercise experiences. Some of the cycling trails and running routes are designed with various public facilities, including showers, drinking fountains, charging docks, changing rooms, and storage rooms (Figure 7-48, 7-49).

Rainbow basketball court

Location: Paris, France | Area: 15 hectares (37.07 acres)
Date: 2021
Designed by Ill-Studio

Dandenong Railway Linear Park

Location: Melbourne, Australia | Size: 22.5 hectares (55.6 acres)
Duration: 2018
Designed by ASPECT Studios

Wangjiang Station

Location: Shanghai | Area: 0.02 hectares (0.06 acres)
Duration: 2017
Designed by Zhi Zheng Architecture Studio

Summary

This chapter has discussed ecological infrastructure, water resources infrastructure, cultural facilities, and recreational facilities within the context of sustainable design. Real case studies have offered valuable insights into the fundamental components of landscape infrastructure and the intricate dynamics within urban spaces and sites. These case studies provide creative ideas for students to engage with infrastructure in their projects, shaping their understanding of the multifaceted role of landscape infrastructure in crafting sustainable, vibrant urban environments.

课后作业
After-class exercises

讨论题
Discussion

景观基础设施设计可以是宏观的生态规划，也可以是日常体验的户外饮水机和休闲座椅。结合户外公共空间的体验，讨论你体验过最好和最不好的基础设施设计，并阐述判断的理由。

Landscape infrastructure design can involve macro-ecological planning, but it can also relate to the everyday experience of drinking fountains and outdoor seating. Discuss one of the best and worst infrastructure designs you have encountered in terms of outdoor public spaces. Provide reasons for your assessment.

设计实践
Design practice

"设计就是构成的过程，也是构筑形式的创造。"请选择案例中的一个景观设施，对它的色彩、材料、功能、构造形式进行设计分析。

"Design is the process of composition and the creation of constructed forms." Select one landscape feature and analyse its colour, material, function, and structure.

推荐读物
Reference reading

1. 冯璐：《弹性景观：风暴潮适应性景观基础设施》，东南大学出版社，2020 年。
2. 韩林桅、张淼、石龙宇：《生态基础设施的定义、内涵及其服务能力研究进展》，《生态学报》2019 年第 19 期。
3. 何志森、杨薇芬：《大地之上：基于人的尺度的图绘》，《国际城市规划》2019 年第 6 期。
4. Cheshmehzangi, A. *Green Infrastructure in Chinese Cities.* Springer, 2022.
5. Firehock, K. E. and Walker, R. A. *Green Infrastructure: Map and Plan the Natural World with GIS.* Esri Press, 2019.
6. Nakamura, F. *Green Infrastructure and Climate Change Adaptation: Function, Implementation and Governance.* Springer Nature, 2022.
7. SWA 基础设施研究提案编：《景观基础设施：SWA 事务所作品分析（原著第二版）》，曾颖译，中国建筑工业出版社，2014 年。

第八章
环境挑战与设计未来

Chapter 8
Environmental challenges and the future of design

学习要点
Study Points

全面了解景观设计面临的全球环境危机，明确专业责任和职业道德要求。

To gain a comprehensive understanding of the global environmental crisis confronting the field of landscape design and to clarify the professional responsibilities and ethical requirements inherent to this discipline

学习目的
Teaching Objectives

通过前沿的学术议题和设计实践，对学科研究动态产生初步的认知，了解先进的研究理论、观点、实践工具和解决问题的方法。

Through cutting-edge academic topics and design practice, students will have a preliminary knowledge of the discipline's research dynamics and acquire advanced research theories, diverse perspectives, and valuable tools that are indispensable for solving problems.

本章节通过可持续设计的视角，探讨景观设计与当代人文社会、环境危机、城市问题、可再生材料、数字技术等的交叉连接。通过职业道德、城市危机、材料危机、科技创新这四个方面进行未来景观设计发展趋势的讨论。

This chapter explores the cross-connections between landscape design, contemporary humanities and society, the environmental crisis, urban challenges, renewable materials, and digital technology through the lens of sustainable design. The future trends of landscape design are discussed through four aspects: professional ethics, urban crisis, material challenges, and technological innovation.

一、职业道德：景观、权利与民主

查尔斯·维斯特·丘奇曼认为，许多科学家都在忙于用污染、具有伤害性的武器以及对于贫穷饥饿的冷漠，来帮助设计一个不公正的世界。[126]

景观设计师应坚定不移地致力于解决全球环境问题，促进人类社会的福祉，培育丰富的文化和社区，并确保全球生态系统的整体状况。[127]人与自然的和谐发展是景观设计从业者的目标；通过长远有力的设计方法促进经济并承担环境保护的责任，实现历史文化符号的传承和人文关怀，是景观设计专业的核心素养。

联合国在1972年发布的《人类环境宣言》中提出，人类社会需要以可持续发展的新思想进行城市设计的革新。2015年9月的联合国可持续发展峰会上，193个成员国正式通过17个可持续发展目标，旨在通过设计综合解决社会、经济和环境的三个维度的发展问题。可持续发展目标呼吁所有国家（不论该国是贫穷、富裕还是中等收入）行动起来，在促进经济繁荣的同时保护地球。目标指出，消除贫困必须与一系列战略齐头并进，包括促进经济增长，解决教育、卫生、社会保护和就业机会的社会需求，遏制气候变化和保护环境。

景观设计是建立在自然系统和人居环境观察上的学科。环境设计的每一个决策，都体现了职业道德、专业素养、公民权益、区域和平、社会平等的各个方面。面对极端的气候变化和危机四伏的文化冲突，景观设计学科需要肩负更重大的社会责任，使景观空间的设计、保护和管理成为解决人类生存危机、增进人类福祉的重要手段。

8.1 Professional ethics: landscape, rights, and democracy

Charles Wester Churchman argues that many scientists are busy using pollution, injurious weapons, and indifference to poverty and hunger to help design an unjust world.[126]

Landscape architects, in contrast, must steadfastly uphold their commitment to addressing global environmental concerns, fostering the well-being of human society, nurturing the richness of cultures and communities, and ensuring the overall condition of global ecosystems.[127] The overarching objective of landscape architecture professionals is to foster the harmonious development of people and nature. Moreover, landscape architects bear the significant responsibility of upholding economic and environmental protection through a sound and longstanding design approach. They must also possess the capacity to pass on historical and cultural symbols and address humanistic concerns, which constitute fundamental qualities of landscape architects.

In 1972, the United Nations issued the Declaration on the Human Environment, advocating for innovative urban design approaches aligned with the principles of sustainable development. Subsequently, in September 2015, during the United Nations Sustainable Summit in Colombia, attended by 193 member states, 17 sustainable development goals were formally adopted, intending to address the social, economic, and environmental dimensions of development through innovative design strategies. They call for collaborative efforts between developing and affluent nations to advance global economic prosperity while safeguarding the environment. They also emphasize meeting social needs such as education, healthcare, social security, and employment opportunities through innovative design strategies to mitigate the adverse effects of climate change-induced natural disasters.

Landscape design is a discipline deeply rooted in the observation of natural systems and human habitats. Every decision about environmental design strategy reflects aspects of professional ethics, professionalism,

1. 景观设计维护和平的环境

景观设计学的综合应用范畴包括国土面积空间规划、开发模式、建造材料等。其中，最迫切的目标是创造一个和平、包容、可持续的社会环境，这是公平的缩影。在景观设计实践中，空间和平的方式是围绕着优化人与环境的保护和整合而展开的。区域环境的规划必须反映人与自然的和平共处。由于人类不断开发和砍伐，自然环境受到了极大的侵害，从而引起环境污染和极端自然灾害的反噬。另外，要保持城市社区规划的公平和避免暴力威胁。良好的社区环境规划应避免在具有废弃建筑、有害土地或者其他影响人类生存安全的区域进行选址。要为儿童、妇女、老人提供安全的环境。同时，发展公共绿地和环境优美的公园可以缓解焦虑和暴力等负面因素的滋生，为健康稳定

的和平环境提供重要的保护网。

2. 景观设计的公平权益

城市公共空间环境的公平性具体体现在对于不同年龄、性别、种族、宗教、社会背景的群体以及弱势群体在空间上都具有关怀性。对城市社区环境的提升和改善不因贫穷和富有而差别对待。对于住房建筑、社区公共设施、植物种植、水体环境等，应确保设计合理性和规范性。针对性别平等问题，应特别注重为妇女和儿童营造安全的环境，如建立视觉盲点少的场所，并确保安全事件得到重视和妥善的处理。在公共空间的设计上，不仅需要满足老人、儿童对于安全防护措施的使用需求，对于母婴、残疾人等特殊群体也需要在空间上体现环境的关怀，如设计残疾人坡道、第三卫生间、残疾人卫生间、夜间照明、应急救护设施等（图

citizenship, regional peace, and social equity. In the face of climate change and crisis-ridden regional cultural conflicts, landscape architecture programs must emphasize a strong sense of social responsibility. The design, conservation, and management of spatial planning strategies will emerge as a pivotal mission that can yield substantial benefits for humanity.

8.1.1 Landscape design maintains a peaceful environment

Landscape design encompasses a wide array of elements, including spatial planning, development patterns, and innovative construction materials. Among these, the most compelling objective is creating a peaceful, inclusive, and sustainable social environment, which is the epitome of equity. In landscape design practice, the way to "make peace" within a space revolves around optimizing the protection and integration of both people and the environment. The planning of regional environments should inherently reflect a peaceful coexistence between people and nature. In contemporary times, the natural environment has been deeply traumatized by human development and deforestation, resulting in environmental

pollution and catastrophic environmental crises. Effective community environmental planning should avoid areas blighted by abandoned structures, hazardous terrain, or sites detrimental to human health. A well-designed community environment should offer a secure setting for children, women, and the elderly to engage in recreational activities. Simultaneously, the creation of vibrant public green spaces and parks can act as a countermeasure to the negative impacts of anxiety and violence, establishing a vital safety net for a healthy and stable environment.

8.1.2 Equity in landscape design

Equity in the urban public space manifests through a careful consideration of the needs and well-being of individuals of different ages, genders, races, religions, social backgrounds, and disadvantaged groups. The enhancement and improvement of the urban community environment should be characterized by a steadfast commitment to non-discrimination. This principle extends to the design of residential structures, community facilities, landscaping, and water bodies, all of which should adhere to rationality and stringent safety regulations. In response

8–01、8–02）。

3. 景观设计的可持续教育意义

自然环境是天然的教育课堂，通过城市景观空间对自然和人居环境进行保护，让市民学习珍惜环境资源，对从个人开始提升环境保护意识有着重要的教育意义。国际公共空间设计已经展开"非设计专业"的环境认知导览。将城市公园历史的介绍、公共空间更新的生态环境变化，以及城市本土植物和昆虫的认知等活动功能软性植入公共空间。人们在体验城市公共空间的过程中能学习和了解自然和环境知识，从而加强每个人与全球生态环境的联系。

二、城市危机：景观都市主义的全新需求

景观都市主义体现了一种对建筑和城市设计无法创造令人满意的当代城市环境的含蓄批评。[128]

景观都市主义由美国建筑师和城市规划师、前哈佛大学景观设计学系主任查尔斯·瓦尔德海姆提出。瓦尔德海姆认为景观设计在面对城市化的转型时，不仅要关注植物和材料背景的设计联系，更要关注不同行业对城市生活的影响。景观设计师是城市整合规划、设计、建筑、经济、生活方式等的重要媒介。在城市发展进入存量时代的背景下，可开发的土地面积已经达到极限，如何进行城市的再设计是我们必须关注的问题。城市从传统的密集、紧凑的建筑形态转向横向广阔的基础设施模式，这种新的发展形式给予了景观设计学新的定义，这就是景观都市主义主要的内涵概述。景观设计对象不仅仅是公园或者花园，面对城市形态的发展，它必须成为建筑和城市规划的连接媒

to gender inequality, particular emphasis should be placed on fostering a safe environment for women and children, such as creating places that minimize visual blind spots, and ensuring that incidents are not overlooked or unaddressed. In the design of public spaces, compliance with safety standards for the elderly and children, mothers with infants, and individuals with disabilities is imperative. For example, the design should include features such as ramps, washrooms for non-binary people and those with disabilities, nocturnal illumination, and a layout that facilitates swift access to emergency response vehicles (Fig 8-01, 8-02). It is crucial to underscore that the enhancement and improvement of the urban community environment should be uniformly accessible and not contingent upon economic status or disparities.

8.1.3 Sustainability education in landscape design

The natural environment serves as an invaluable classroom. Through the preservation of nature and the human environment in urban landscapes, citizens can gain insights and develop an appreciation for environmental resources. The cultivation of environmental awareness stands as a pivotal undertaking, and one approach to achieving this is through international public space design, which incorporates "non-design professional" into environmental awareness tours. These tours effectively introduce individuals to the history of urban parks, the ecological transformations occurring within public spaces, and the recognition of urban native plants and insects. By engaging with these aspects, the importance of public spaces is underscored. People can gain a deeper understanding of the processes of environmental change by experiencing urban public spaces, thus forging a meaningful connection between each individual and the global ecological environment.

8.2 Urban crisis: a new demand for landscape urbanism

"Landscape urbanism embodies an implicit critique of the inability of architecture and urban design to create a satisfying contemporary urban environment."[128]

Landscape urbanism was pioneered by Charles Waldheim, who served as the director of the Harvard Graduate School

图 8-01 **Multipurpose toilets**
多用途厕所

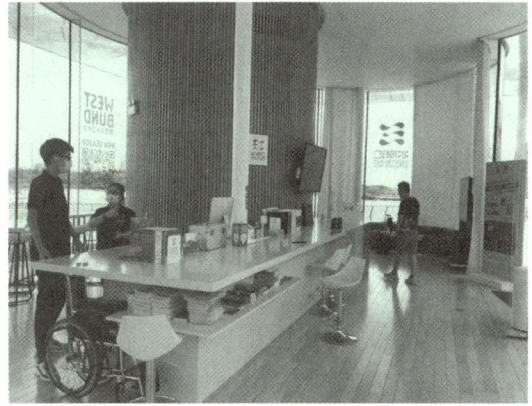

图 8-02 **First aid facilities**
急救设施

of Design. Waldheim asserted that landscape design should not solely revolve around the aesthetics of plants and materials but must also consider the far-reaching impacts of various sectors on urban life. Landscape architects are essential in bridging the realms of urban planning and design, architecture, economics, and the urban lifestyle. As vacant land diminishes in many cities, most urban centres find themselves on the brink of development saturation, necessitating a fundamental rethinking of urban landscapes. The urban development paradigm has shifted from the traditional dense and compact configurations to a horizontal and expansive infrastructure model. This evolution brings about a redefined role of landscape architecture, which emerges as the primary practice within the framework of landscape urbanism. Landscape design transcends the confines of mere parks or gardens; it becomes the integrative medium that binds architecture and urban planning in the context of urban development. The urban design process is an ongoing exercise in balancing human and natural environments to arrive at the most appropriate planning and design decisions. Among them, landscape urbanism is a pragmatic exploration, focusing on three critical aspects in addressing the urban crisis.

8.2.1 Landscape urbanism is the connecting medium of urban design

James Corner, a professor of landscape design at the University of Pennsylvania and a renowned designer of the High Line Park, once said, "Traditional urban design and planning can no longer function effectively in contemporary cities, and this new thinking and action has become inevitable."[129] Conventional urban designs, predominantly executed by architects, have typically involved filling a city's spaces with architectural structures while placing an area of green space as a landscape where no building is needed. As a result, urban design emerged as a novel profession aimed at improving the single-dimensional urban spatial planning carried out by architects. However, it became increasingly evident that this architectural-centric planning strategy often failed to meet the needs for creating inviting and comfortable public spaces in urban environments. Consequently, most modern planning projects require the collaborative efforts of diverse design

介，在城市设计的过程中不断平衡人居环境和自然环境，做出最合适的规划设计。面对城市的危机和问题，景观都市主义主要进行了以下三个方面的实践探索。

1. 景观都市主义是城市设计的连接媒介

美国宾夕法尼亚大学景观设计学教授和著名高线公园设计师詹姆斯·科纳曾表示，设计与规划已经无法在当代城市有效运转，这种全新的思考和行动已成必然。[129] 建筑师进行城市设计时，习惯于用建筑的形式去填充城市的空间，而在不需要建筑的地方则放置一片绿地作为景观的区域。为了改善建筑师对于城市空间单一视角的规划方式，城市设计作为新的专业应运而生。然而，越来越明显的是，这种以建筑为中心的规划策略往往不能满足在城市环境中创造吸引人和舒适的公共空间的需要。因此，大多数现代规划项目需要不同设计团队协作努力。负责城市规划的团队必须包括一系列的专业人士，包括建筑师、规划师、景观设计师、政府官员、社区居民和其他利益相关者。

19 世纪初期，美国底特律经历了大规模的城市建设，以密集的房屋和紧凑的城市空间规划满足城市激增的人口。但城市在经历数十年的产业改革后，主要的娱乐产业投入和人口的流失造成底特律政府入不敷出的财政问题，从而成为美国最大的宣告破产的城市。衰败的经济环境同时带来了灾难性的空置率、废弃建筑和社会冲突（图 8-03）。20 世纪初，丹·霍夫曼推行拆除建设的理念，为底特律提出拆除工程，把大面积的建筑空间改善为城市绿地和公园，从而缓解了底特律的公共空间矛盾。景观设计对于城市空间不仅仅是绿化植物的设

teams. The team responsible for the urban planning process must encompass a spectrum of expertise, including architects, planners, landscape architects, government officials, community residents, and other stakeholders. In the early nineteenth century, the city of Detroit in the United States experienced a period of intense urban development, characterized by dense housing and compact spatial planning, an imperative response to the city's burgeoning population. However, decades of industrial changes, coupled with failed investment in major entertainment industries and a decline in population, plunged Detroit into a financial crisis, ultimately resulting in it becoming the largest US city to declare bankruptcy. The deteriorating economic environment further precipitated alarming urban challenges, marked by abandoned buildings, soaring urban vacancy rates, and socially contentious living conditions (Fig 8-03). It was not until the early twentieth century that Dan Hoffman championed the concept of demolition and reconstruction, proposing large-scale demolition projects to transform vast areas of built space into urban green spaces and parks. This endeavour alleviated Detroit's public space problems.

The allure of landscape design for urban spaces extends beyond the mere inclusion of greenery and plants; it represents a comprehensive expression of the place's architecture, texture, colour, and culture. Landscape design possesses the unique ability to reinstate spatial form and social order. Furthermore, it introduces novel modes of social interaction. Effective urban design necessitates the implementation of measures involving re-orientation, slowing down, and prohibition.

8.2.2 Landscape urbanism proposes new requirements for sustainable urban design development

The transition from landscape urbanism to the concept of urban acupuncture[130] is a new exploration of design development in the face of urban space and development crises. One of the core issues is thinking about how to maintain the city as time passes, and how to develop healthier and more sustainable spaces to meet the needs of the residents and mitigate against natural disasters. Determined to innovate urban space, people begin to think about how to reuse buildings and renew old neighborhoods to improve the urban public space. The transformation

计，更是对场所建筑、肌理、色彩、文化、精神娱乐方式的综合表现。景观设计具有很强的修复空间形式和社会秩序的独特能力，同时也作为一种新的社交方式出现。良好的城市设计意味着对城市活动和生态进行重新导向、放缓或禁止。

2. 景观都市主义提出城市可持续设计发展的新要求

从景观都市主义到"都市针灸"[130]，都是城市空间发展危机下的新设计探索。其核心问题是如何让城市在衰老的过程中，更健康、更可持续地进行空间的发展，以应对人居环境的需求和自然灾害。城市空间的革新开始重新通过建筑和老旧街区的更新来改善城市公共空间的需求。老旧社区单调的非机动车停车区与展览空间相结合（图8-04），更新改造街道修

复社区环境，或者用雕塑的设计重塑城市公共空间的活力等，都是"都市针灸"理念下进行的非建筑化的改造方式。这种改造方式是景观都市主义的产物，目标是提醒设计师和政府官员以更为节约、更符合人们生活的视角去进行城市环境的改善。

3. 景观都市主义从静态图像转化为都市生活模式的探索

景观都市主义对城市设计模式的影响远比对城市外观的影响更为重要。

风景画般的景观图像已经日益转换为生活场景。当代的景观设计模糊了建筑和绿化区域的边界，形成了联动的整体。从美国的佩雷公园（1967年），到中国上海的永嘉路口袋广场（2021年），城市的景观设计已经从形式上对自然的崇拜和模仿，走向对人居生活模式

of industrial districts with old factories into children's playgrounds, the restoration of the residential neighborhood through street renewal (Fig 8-04), and the rejuvenation of urban public spaces with sculptures are all non-architectural methods of transformation under the concept of urban acupuncture. This type of transformation is a product of landscape urbanism, and its goal is to remind designers and government officials to improve the urban environment from a more economical and human perspective.

8.2.3 The exploration of landscape urbanism's transformation from images to urban living patterns

The influence of landscape urbanism on the evolution of urban design patterns transcends mere aesthetics, emphasizing a deeper, more profound impact.

Landscape-like Images are transformed into living and breathing urban scenes. In contemporary landscape design, the once-clear boundaries between architecture and green spaces have blurred, yielding a more interconnected fusion. From the inception of the US urban park Perry Park (1967), to the recent creation design of Yongjia Road

Square in Shanghai, China (2021), urban green spaces have progressed from formal representations and imitation of nature to a profound concern for actual living patterns. In the design of Perry Park, the intrinsic natural elements, including waterfalls and tree formations, have been thoughtfully preserved. However, it has been reimagined as a corner square nestled amidst towering office buildings, and the arrangement of party-style seating represents a novel exploration of urban social dynamics. On the other side of the globe, China's Yongjia Road Plaza breathed new life into a disused and bleak building occupying a prime location, reimagining it as a vibrant community square. The introduction of resting lanes injects renewed vitality into the urban space (Fig 8-05). The plaza lounge has swiftly emerged as a bustling hub for social activities, media displays, and vibrant neighborhood life.

8.3 Landscape regeneration: renewable resources and design

"Without landscape regeneration, there is no landscape sustainability. Fostering the regeneration and self-renewal capacity of healthy landscapes and restoring the

的关注。佩雷公园的设计保留了瀑布和树阵的自然元素，给予城市自然环境的气氛。但规划定位在密集的办公楼附近成为街角广场，还布置了聚会式的座椅，都是对城市社交模式的新探索。永嘉路口袋广场通过改造徐汇区的一处废弃空地，把黄金地段的建筑空地改造为社区广场。通过休息连廊的置入，重新赋予城市空间活力（图 8-05）。休闲广场成为该片区社交活动、媒体展示、街区生活的重要载体。

三、景观再生：可再生资源设计

"没有景观的再生，就没有景观的可持续。培育健康景观的再生和自我更新能力，恢复大量被破坏的景观的再生和自我更新能力，便是可持续景观设计的核心内容，也是景观设计学的根本的专业目标。" [131]

城市建造资源的浪费加剧了全球环境危机。在有限的资源存量下，人居环境的建造和环境设计的发展，必须面向可再生资源的设计。

可再生资源是指能够通过自然力以某一增长率保持或增加蕴藏量的自然资源。可再生资源设计主要是从材料的选择和使用率方面进行设计，从而实现资源的持续利用。可再生资源的持续利用可保证资源的使用不受自然存量的限制。风能、水能、核能、太阳能、热能、生物质能等清洁能源都可转化并运用到城市的各种耗能中。另外，对于材料的回收再利用也是可再生资源的转换方式，可以减少过度开发和浪费带来的环境污染。

可持续再生的过程不仅局限于材料部分的再生，也注重景观系统的全生命周期设计。从项目选址和施工技术，到材料选择、项目管理

regeneration and self-renewal capacity of a large number of destroyed landscapes are the core elements of sustainable landscape design, which is also the fundamental professional goal of landscape design." [131]

The wanton depletion of urban construction resources has led to a severe global environmental crisis. In the face of finite resources, the development of construction and environmental design must pivot towards the utilization of sustainable and renewable resources. Renewable resources, being bountiful in nature and replenished at a consistent rate by natural forces, emerge as ideal candidates for material selection, thus ensuring the sustainable use of resources. This approach safeguards resources against being exhausted by natural supply constraints. Renewable resources encompass, among other things, the harnessing of clean energy sources such as wind, water, nuclear, solar, thermal, and biomass for meeting urban energy needs. On the other hand, the recycling of materials stands as a prime example of renewable resource utilization, mitigating environmental pollution stemming from excessive exploitation and waste.

The concept of sustainable recycling extends beyond the material part of the recycling and encapsulates a holistic life cycle design for the landscape system. Every facet, from project site selection and construction techniques to material choices, project management operations, and allocation of people and resources, can collectively culminate in the creation of sustainable resource design. The Leadership in Energy & Environmental Design Building Rating System (LEED), implemented by the US Green Building Council, is internationally recognized as the most comprehensive and influential rating standard. It is divided into diverse green design rating criteria tailored to distinct project types. The eight principal landscape, architectural, and planning projects are as follows.

- LEED BD+C Building Design and Construction
- LEED ID+C Interior Design and Construction
- LEED O+M Operations and Maintenance
- LEED for Cities and Communities Cities and Communities
- LEED for Residential Housing
- LEED for Transit Stations
- LEED Zero Near Zero Building
- LEED ND Community Development

运作、人力和资源配置，每个方面都可以共同践行可持续的资源设计。通过可持续设计，对可再生资源、人文景观和健康生产力建立可持续的网络链接。美国绿色建筑委员会推行的《绿色建筑评估体系》（Leadership in Energy and Environmental Design，简称 LEED）是国际上最为完善和最具影响力的评估标准之一。其中，针对不同的项目类型制定了不同的绿色设计评定标准，涵盖八大景观、建筑、规划项目，具体如下。

· LEED BD+C 建筑设计施工
· LEED ID+C 室内设计与施工
· LEED O+M 运营与维护
· LEED for Cities and Communities 城市与社区
· LEED for Residential 住宅
· LEED for Transit 交通站点
· LEED Zero 近零建筑
· LEED ND 社区开发

每个项目的评价指标都列有详细的可持续设计条件、材料选择、技术标准，以及管理运营的方法，为不同类型项目进行可持续设计提供全周期的参考（图 8-06）。

景观设计对于可再生资源的运用，主要集中在三个方面。

1. 减少损失和浪费

在景观设计的过程中结合利用自然中的水、光、热、风等清洁能源进行合理的场地规划和建筑设计，从而避免过多的耗能和浪费。例如，选择常年光照范围较大的地方作为主要活动空间，通过依循自然光照和流通风向来减少室内灯光及空调的耗能。在城市绿化种植中，使用本土物种和低维护的植物配置，可以节约

Detailed sustainable design conditions, material selection, technical criteria, and methods for managing operations are listed for each project's evaluation metrics. The full-cycle design reference for sustainable design for various project types is shown in Figure 8-06. The integration of renewable resources into landscape design is cantered around three primary areas.

8.3.1 Reduce loss and waste

In the landscape design process, natural resources such as water, light, heat, and wind, along with other clean energy, play a pivotal role in rational site planning and architectural design. This approach effectively mitigates excessive energy consumption. For example, areas abundant in year-round sunlight can be strategically designated as the primary office and activity spaces to reduce indoor air conditioning demands through natural light utilization and natural ventilation. When it comes to urban greenery planting, the deployment of low-maintenance plants and the preference for native flora serve multiple purposes such as conserving water, reducing the use of fertilizers and herbicide chemicals, and safeguarding the ecological balance of the soil.

A notable illustration of this sustainable design practice can be found in the California Academy of Sciences, where a multitude of recycled natural resources is harnessed to curtail the building's energy requirements. In particular, the building boasts an extensive "eco-roof" installation designed to reduce indoor air-conditioning usage (Fig 8-07). In addition, a transparent glass roof, capable of breathing, incorporates photovoltaic cells, contributing 5 percent of the museum's total electricity generation. The project's sustainable design revolves around the symbiotic use of natural light, rainwater harvesting, and self-sustained energy generation. While the beautiful landscape of the city and its buildings are integral, it is equally imperative to contemplate long-term maintenance strategies and associated costs during the design process. This forward-thinking approach is where designers must channel their attention and focus.

8.3.2 Material recycling and utilization

The utilization of recycled resources, ranging from wasteland and household waste to old building structures, steel, and masonry, yields notable benefits by substantially

图 8-03　The empty city of Detroit
底特律的空城

图 8-04　Community renewal
乳山新村地下博物馆社区更新

图 8-05　Pocket Square, Yongjia Road, Shanghai
上海永嘉路口袋广场

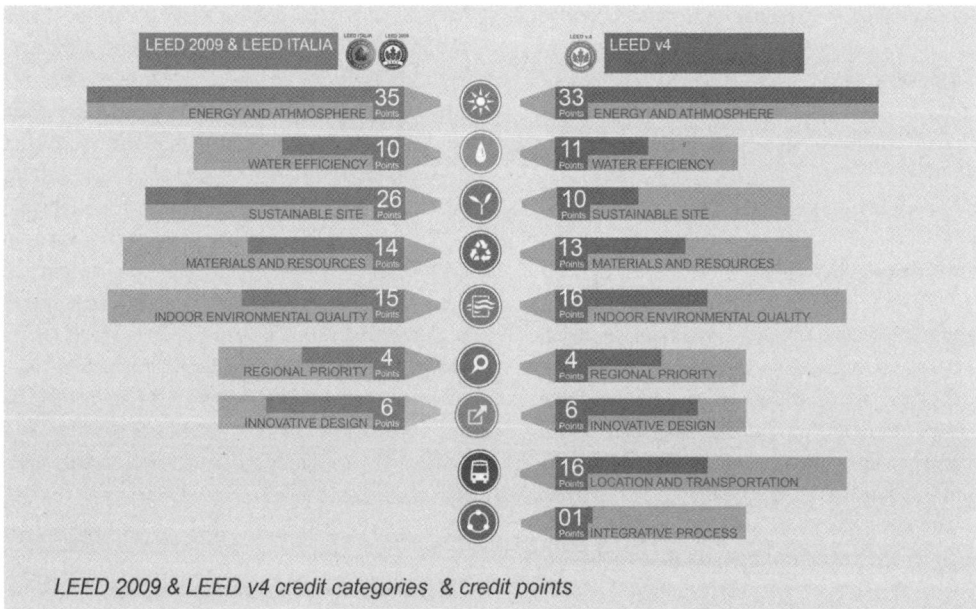

图 8-06　LEED rating system
LEED 评估标准

灌溉用水，减少化肥和除草剂等化学品的使用，保持土壤生态不受破坏。

美国加州科学院的建筑设计就应用了多种可循环自然资源来降低建筑的耗能需求。包括运用大面积的"生态屋顶"作为建筑的屋面来降低室内对空调的需求（图8-07）；通过"可呼吸"的透明玻璃顶打造光电池，为建筑提供5%的电力；使用自然采光、雨水收集、自发能源等多种清洁能源组合的可持续设计。城市优美的环境和建筑需要持续的运用和维护，因此设计过程中设计师应重点关注未来维护的方式和成本。

2. 材料再回收和利用

对废弃土地、生活废品、旧建筑结构、钢材、砖石等进行重新利用，可大幅减少建筑耗材，降低建造过程中对环境的重复污染。面对

全球资源的匮乏和各国环境保护意识的提升，旧建筑的更新设计成为可持续设计方法的重要方面。旧有建筑或者废弃厂房拥有较完善的框架基础，并且保留有价值的建筑形态对于保护城市历史文化具有重要意义。美国纽约的高线公园保留了旧有高架铁路的形态，成为城市中最受欢迎的公共空间之一。它不仅减少了拆除高架铁路的巨额施工费用和对环境的污染，还非常巧妙地结合了高架铁路的历史文化，唤起了市民对于当时滨水码头货运繁忙场面的历史记忆。中国上海的龙美术馆，也是在保留滨水码头煤料斗卸载桥和运货轨道设施的基础上，发展出独特的工业感造型（图8-08）。这种颇具前瞻性的方法是当代设计师必须关注的地方。

curbing the cost of construction materials and limiting environmental pollution during the construction process. Given the current global resource scarcity and the growing emphasis on environmental protection, the renewal design of ageing structures emerges as an essential part of sustainable design strategies. Old buildings and abandoned factories possessing solid frameworks and architectural significance retain intrinsic value in the context of urban history and culture. A prominent example of this approach can be witnessed in New York City's High Line Park, which has burgeoned into one of the city's most beloved public spaces by preserving the original form of the old elevated railroad. This conservation effort not only obviated the need for a costly dismantling process but also mitigated the environmental ramifications that would have accompanied such deconstruction. It cleverly weaves together the historical essence of the elevated railroad, evoking the prosperous scene of the waterfront docks for freight transportation. Across the globe, in Shanghai, the Long Museum of Art has masterfully adopted a unique industrial design concept by retaining the coal hopper unloading bridge and waterfront pier's freight rail infrastructure

(Fig 8-08). Recycling and repurposing materials provide designers with an alternative canvas to unleash their creative potential.

8.3.3 Regenerative design

The natural growth process follows a cyclical pattern of germination, fruiting, withering, and decay. When plants wither, they transform into nutrients for the soil, nourishing the growth of new plants. In nature, nothing goes to waste. The principles of this cyclical system find application in urban construction as well. The concept of reusing waste, sewage, and contaminated soil through material recycling and reclamation to obtain new recycled materials is integral to this paradigm. Within the built environment, a wide array of components, including structures, walls, paint, rubber, and paving, can be replaced with new, low-consumption building materials that are developed from recycled resources. For example, traditional paving materials use recycled plastics to reduce the heat generated during asphalt construction, resulting in a nearly 20 percent reduction in fuel consumption. Recycled packaging, plastic bottles, and various plastic products can be repurposed into landscape

3. 再生设计

自然中有生长、结果、凋谢、消解的循环过程。植物凋谢以后成为土壤的养分，从而滋养新的植物生长。所以在自然界中没有"垃圾"的概念。景观设计试图在城市建设中运用同样的理念，即通过材料的回收和再生，对垃圾、污水、被污染的土壤进行重新利用，从而获得新的再生材料。建造环境中的结构、墙体、涂料、橡胶、铺装等都可以用新的、低消耗的建筑材料来替代，这些材料都是从回收资源中开发出来的。例如，如果用再生塑料替代传统的铺路材料，可以减少沥青施工过程中产生的热量，从而减少近20%的燃料消耗。再生包装、塑料瓶等各种塑料制品可以重新用作景观铺地材料。即使是用塑料和橡胶垃圾制成的砖，也可以成为传统墙壁的可靠替代品，减少建筑项

目对环境的影响。可再生砖[132]在世界多个经济适用性房和紧急避难所项目中得到应用（图8-09）。

四、科技创新：数字化景观设计

数字技术的发展带动了人工智能、虚拟现实等领域的发展，让城市的研究、规划、体验产生全新的变化。数字技术与国土空间规划、景观设计的结合，令传统的设计研究、设计体验、设计元素、设计手法产生了更立体、更直接、参与感更强的改变。[133]

1. 信息化技术辅助设计判断

景观专业的学生对于通过数字技术来了解城市空间环境已经不陌生，常使用计算机图纸绘制、三维建模、地理信息查找等方式进行设计实践。随着数字技术的拓展，越来越多的地

paving materials. Even bricks made from discarded plastic scrap and rubber become reliable alternatives to traditional walls, reducing the environmental footprint of construction projects. These recycled bricks have found application in several affordable housing initiatives and emergency shelters worldwide (Fig 8-09).

8.4 Technology innovation: digital landscape design

The advancement of digital technology has given rise to the development of artificial intelligence and virtual reality technology, which have fundamentally transformed the urban planning process, affecting research, development, and overall experiential aspects.[133] By harnessing the potential of digital technology and fusing it with territorial spatial planning and landscape design, traditional design research, design experience, design elements, and design methodologies have evolved into a three-dimensional, direct, and participatory way.[132]

8.4.1 Information technology–assisted design

Landscape students are no strangers to leveraging data-driven technology to comprehend the urban spatial

environment. This is evident in computer-assisted design practice, including drawing, 3D modelling, and the retrieval of geographic information. However, as digital technology continues to expand, a large number of geographic information systems are now delivering a massive amount of data through the Internet and big data platforms to facilitate exploration from diverse design perspectives. Notable examples include GIS (spatial information platform), POI, SuperMap Objects, and Visual Basic 6.0 programs, which serve as tools for geographic information management, map management, query tracking, and network analysis through digital technology. These tools allow designers to transcend temporal and spatial limitations and analyse various environmental features. For example, the use of GIS and POI regional data analysis can aid in the allocation of public sports facilities' space in Beijing, China. This approach enhances the precision of addressing design issues in the early stages of planning for sports and recreation activities.

In addition, digital technologies provide simulations for terrains with complex topography and problematic

图 8-07
Green roofs
生态屋顶

图 8-08
Long Art Museum
龙美术馆

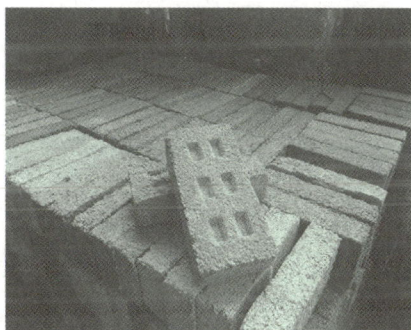

图 8-09
Renewable brick
可再生砖

理信息系统通过互联网和大数据进行信息空间呈现，以满足设计视角的不同探索。例如，GIS、POI、SuperMap Objects、Visual Basic 6.0 等都是通过数字技术进行地理信息管理、地图管理、查询跟踪、网络分析的方式，让设计师能够突破时间和空间的限制，对不同的环境特征进行分析。例如，使用 GIS 和 POI 区域数据可以分析得出中国北京公共体育设施空间的分布情况，从而在体育休闲空间规划前期更精准地聚焦设计问题。

另外，数字技术可以提供全过程的建造模拟，为地形复杂、施工难度极大的地形环境呈现和测评设计方案。如通过 BIM（Building Information Modeling，建筑信息模型）的环境模拟，对建造环境中出现的设计问题进行检查并模拟解决方案，可以积极规避施工过程中因

判断失误而造成的生命和财产损失。自然景观环境也可以通过数字化模拟场景进行渲染，通过电脑或者 VR 设备体验设计后的环境气氛（图 8-10），为设计师感受空间营造效果提供更直接、更快捷的方式，同时为项目推广和汇报提供有效的支持材料。

2. 数字技术拓宽对环境的认知

数字技术开拓城市智慧化管理。通过数字技术可以进行能量损耗、灯光控制、温度调节、场所人流动向等方面的管理，还可以进行建造模拟测试，为人类探索空间改造提供全面的模拟平台。智慧城市、智慧公园、智能化管理系统可以为城市的公共空间提供更精准和人性化的服务。[134]

智慧城市中，通过城市数据的输入和空间模型的搭建，可为城市的规划设计、建筑施工、

construction scenarios, notably through the use of BIM (Building Information Modelling) environment simulation. By simulating the building environment with the BIM model, designers can evaluate and simulate solutions for construction-related design challenges. This helps avert economic and human losses that might arise from judgment errors during the construction process. On the other hand, digital simulation can effectively convey the natural landscape environment, allowing people to experience the ambiance of the designed environment through computers or virtual reality equipment (See Figure 8-10). This approach provides designers with a more immediate and expeditious means to gauge the impact of spatial creation while generating compelling materials to support project presentations and reports.

8.4.2 Digital technology broadens understanding of the environment

Digital technology ushered in a new era of intelligent urban management, offering a multitude of applications for energy conservation, lighting control, temperature regulation, traffic flow, and construction simulations. They

also provide a robust simulation platform for exploring spatial transformation. Smart cities, smart parks, and intelligent management systems collectively enhance the precision and human-centric nature of public space services within urban environments.[134]

By integrating urban data and constructing spatial models, Smart Cities can provide visual representations for urban planning and design, building construction, urban operations, park management, and asset operation and maintenance. For instance, a city data platform can be used to identify areas prone to traffic congestion and facilitate traffic management and diversion strategies. In addition, it can provide insights into electronic irrigation and pollution measurement, monitoring urban soil moisture and greenery quality. Therefore, these applications significantly reduce the costs associated with urban greening management.

Another example is the smart park, which integrates smart facilities into urban green space. These facilities encompass functions such as monitoring the park's pedestrian flow,

城市运营、园区管理、资产运维等展示可视化的数据。例如，城市数据平台可用于识别容易发生交通拥堵的区域，促进交通管理和分流策略。此外，还可以为电子灌溉、污染测量、城市土壤水分、绿化质量监测提供参考。这些应用将大大降低城市绿化管理相关的成本。

智慧公园是对城市绿地智慧化设施的统筹设计。监测公园的行人流量、跟踪行人体验和心率的变化、指导植物的观赏开花期，以及根据环境温度调整表面等，都是数字技术优化空间体验场景的方法。智慧公园的设计，可以拓展公园休闲体验的维度，为人们的生活出行提供更为便利的辅助设施。智慧服务是智能化管理的体现，如帮助个人寻找厕所和饮水机的云技术，以及为公共空间应急服务量身定制的导航系统。[135] 人们可以通过手机端进行信息的

查找，从而体验更友好的服务环境。

3. 数字技术突破地域的空间体验

自然元素是景观设计中重要的构成元素，如果没有自然的植被、水体、光线等，那么它还是自然景观吗？这个问题让我们开始思考，面对数字化的技术，我们应该如何处理自然环境和虚拟环境的关系，并了解使用数字技术对我们理解环境有什么帮助。

数字技术给空间的体验提供了全新的视角[136]，超越了地理和时间的界限，允许个人在不离开家的情况下环游世界。例如，谷歌地球提供了一个平台，可以通过 3D 视图体验世界各地的城市社区，不仅能向人们介绍不熟悉的城市和环境，还能提供精确的导航。人们不仅能进入模拟的城市环境，还可以通过谷歌的博物馆信息平台进入虚拟的世界博物馆丰富展

tracking changes in the pedestrian experience and heart rate, offering guidance on the ornamental flowering periods of plants, and adjusting surfaces to match ambient temperatures. Each of these features represents an application of digital technology aimed at optimizing spatial experiences. Smart park design broadens the dimensions of the park's leisure experience and provides more convenient auxiliary facilities for daily life. Knowledge service is the embodiment of smart management, such as cloud technologies that assist individuals in locating restrooms and water fountains, along with wayfinding guide systems tailored to emergency services in public spaces.[135] This enables people to access information through their mobile devices.

8.4.3 Digital technology breaks through the spatial experience of the territory

Natural elements are important in landscape design. Without green plants, water bodies, or natural light, a landscape hardly retains its natural essence. In the context of digital technology, city planners carefully contemplate how to deal with the relationship between the natural and

virtual environments and harness digital technology to foster a deeper appreciation of the natural environment.

Digital technologies offer a new perspective on spatial experiences[136], transcending geographical and temporal boundaries, and allowing individuals to traverse around the world without leaving their homes. For instance, Google Earth provides a platform for experiencing 3D representations of neighborhoods in cities worldwide. It not only introduces people to unfamiliar cities and urban environments but also offers precise navigation. Moreover, Google's museum information platform grants access not only to simulated urban settings but also to the virtual world of museums, enriching exhibition experiences. This confers substantial educational value to visitors while enhancing the dissemination and immersion in global cultural information. It eliminates the need for expensive travel costs and ensures that individuals with limited mobility have the same opportunity to explore the world (Fig 8-11).

Virtual reality technology also expands the possibilities of artistic expression, converting static design images into

览体验，不仅具有实质性的教育价值，而且加强了全球文化信息的传播和沉浸体验。这样的体验方式不需要高额的旅行费用，也为行动不便的人士提供了感受世界的可能（图8-11）。

　　数字化虚拟技术也拓展了艺术表达的可能，将静态的设计图像转化为动态的体验旅程。数字化设计公司Team Lab通过动态模拟的方式，将不同景观场景运用到空间体验上。体验者犹如进入新的时空，在这里数字视觉转换展开，创造出一种与日常生活不同的感官体验（图8-12）。在韩国，被称为"波"的虚拟技术装置通过韩国时代广场的多媒体设计为公共空间注入生命，创造了一个迷人的虚拟景观场景（图8-13）。同时，虚拟技术还可以与文化艺术相结合。例如，北京世界园艺博览会将《千里江山图》和视觉影像相结合，形成了流动的千里江山图景。

　　数字化信息技术为城市景观空间提供了更有互动感和参与感的设计方法和体验方式，这为未来城市智慧化空间的需求和管理建立了重要的技术平台和研究基础。

小结

　　本章节通过介绍景观设计学所面临的环境挑战和城市问题，让学生对所学专业职业道德和责任挑战有一定认知，并且通过对可持续设计和数字技术具体实践的介绍，展现未来景观设计发展的新趋势。

Summary

This chapter provides students with an understanding of the level of professionalism demanded in their field of study. It accomplishes this by delving into the environmental challenges and urban issues that landscape architecture grapples with, while also emphasizing the importance of professional ethics and responsibility. It also offers insights into emerging trends that will shape the future development of landscape design through the creation of sustainable design methodologies and the tangible application of digital technology.

dynamic experiential journeys. Digital design companies, like Team Lab, design various landscape scenarios that can be experienced through dynamic simulations. The experience feels like stepping into a new realm, where digital visual transformations unfurl, creating a sensation that deviates from everyday life (Fig 8-12). In Korea, the virtual technology installation known as "wave" breathes life into public spaces through multimedia design at Korea Times Square, creating a mesmerizing virtual landscape scene (Fig 8-13). At the same time, virtual technology seamlessly intertwines with cultural and artistic content. A prime example is the Beijing World Horticultural Expo, which combines the painting, "Thousand Miles of Rivers and Mountains," with virtual technology to create a flowing representation of this famous Chinese landscape theme.

Digital information technology provides a more interactive and participatory approach to the design and experience of urban landscape spaces. This serves as a foundational technological platform and research bedrock for future urban smart space requirements and management.

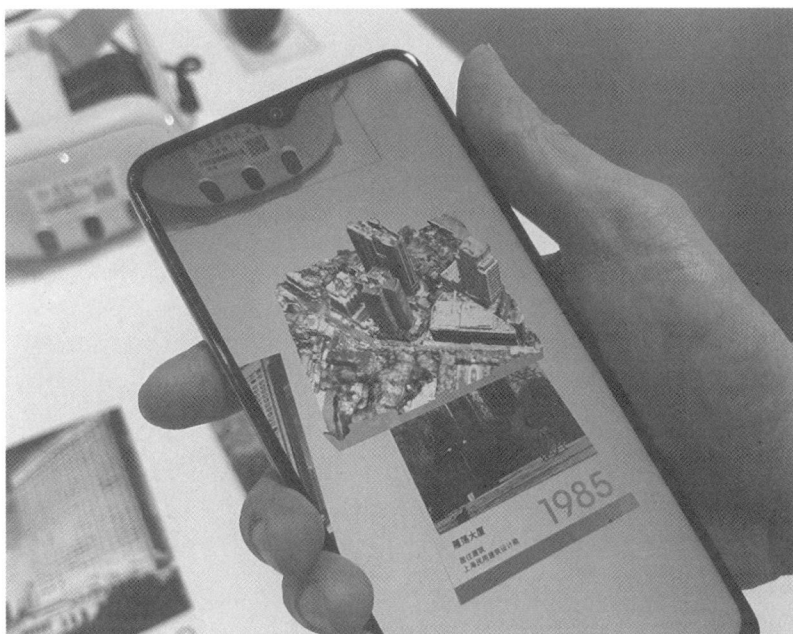

VR scene

图 8-10　VR 场景

Google Museum

图 8-11　谷歌线上博物馆

Team Lab digital design

图 8-12　Team Lab 数字设计

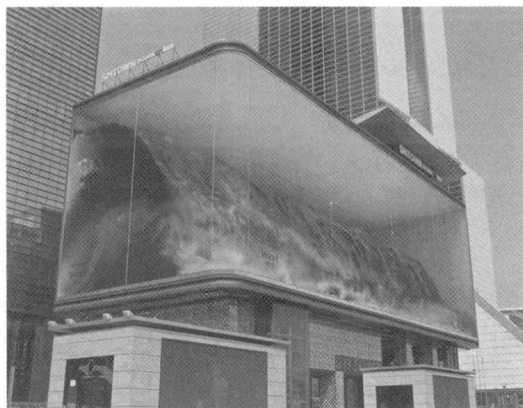

"Wave" on digital billboard

图 8-13　数字广告牌上的"波"

课后作业
After-class exercises

讨论题
Discussion

鉴于当代城市地区面临的环境挑战，你认为需要解决的最紧迫的环境问题是什么？

Given the environmental challenges facing contemporary urban areas, what is the most pressing environmental problem that needs to be solved?

设计实践
Design Practice

根据讨论题内容，请畅想当代景观设计学的追求与使命。

Considering the content of the reflection questions, please think about the pursuit of the mission within the realm of contemporary landscape design science.

推荐读物
Reference reading

1. [美] 查尔斯·瓦尔德海姆：《景观都市主义：从起源到演变》，2018 年。
2. 成玉宁：《数字景观》，东南大学出版社，2019 年。
3. 何志森、杨薇芬：《大地之上：基于人的尺度的图绘》，《国际城市规划》2019 年第 6 期。
4. [荷] 雷姆·库哈斯：《癫狂的纽约》，唐克扬译，三联书店，2015 年。
5. Corner, J. and Hirsch, A. B. ed. *The landscape imagination: Collected Essays of James Corner 1990-2010*. 2014.
6. Paez, R. Operative Mapping: *The Use of Maps as a Design Tool*. Actar, 2019.
7. Waldheim, C. *Landscape as Urbanism: A General Theory*. Princeton University Press, 2016.

图片来源
Image credit

1-01
https://www.thelocal.fr/20220511/in-pictures-see-how-paris-plans-to-transform-the-champs-elysees/

1-03
https://www.mayalinstudio.com/

1-06
https://www.downtoearth.org.in/news/natural-disasters/desertification-in-africa-10-things-you-must-know-54430

2-02
https://en.wikipedia.org/wiki/Emerald_Necklace

2-03
https://en.wikipedia.org/wiki/Denmark

2-04
https://www.archdaily.com/631845/4-techniques-cold-climate-cities-can-use-to-make-the-most-of-their-waterfronts

2-05
https://www.trip.com/blog/6-places-to-feel-the-culture-and-history-of-harbin/

2-06
https://www.jma.co.jp/en/works/urban-planning-for-osaki-gotanda/

2-07
https://capla.arizona.edu/studio/lecture-recap-jie-hu-urban-design

2-08
https://www.bois-petrifie.org/en/the-m-zab.php

2-09
https://tibooburra.org.au/gallery/

2-10
https://thorpetrees.com/advice/table-of-latin-common-names/

2-11
https://www.pwpla.com/projects/saitama-sky-forest-plaza

2-14
https://www.culturedmag.com/article/2021/09/24/piet-oudolf-gardens-are-designed-to-move-you

3-01
https://en.wikipedia.org/wiki/Parkroyal_Collection_Pickering

3-02
https://commons.wikimedia.org/wiki/Category:Pla%C3%A7a_de_Sant_Jaume

3-03
https://www.totbarcelona.cat/politica/maragall-exigeix-colau-reformar-via-laietana-ple-ciutat-58232/

3-04
https://www.lonelyplanet.com/spain/barcelona/la-rambla-and-barri-gotic/attractions/la-rambla/a/poi-sig/1105382/1320690

3-05
https://www.sohu.com/a/539034231_121196329

3-06
https://www.popsci.com/technology/article/2012-07/how-construct-lightest-most-open-olympic-stadium-ever-built/

3-07
https://collection.sina.cn/yejie/2017-03-28/detail-ifycstww1494329.d.html

3-08

https://fr.wikipedia.org/wiki/Unit%C3%A9_
d%27habitation

3-09

https://www.bcn-advisors.com/una-guia-del-
emblematico-barrio-del-eixample

3-10

https://www.mykonosceramica.com/en/the-most-
impressive-urban-layouts-from-an-aerial-view-by-
mykonosceramica/

4-01

https://hoteltechreport.com/news/hotel-entrance-
how-to-make-it-truly-welcoming

4-02

https://www.archdaily.com/286223/superkilen-
topotek-1-big-architects-superflex

4-03

https://www.architonic.com/it/project/topotek-1-
superkilen/20214046

4-07

https://www.amusingplanet.com/2015/12/the-ginkgo-
avenue-in-tokyo.html

4-08

https://www.college.columbia.edu/cct/issue/
springsummer-2021/article/your-favorite-campus-
places

4-09

https://www.alburycity.nsw.gov.au/leisure/arts-and-
culture/annual-events/music

4-10

https://www.aquascapeconstruction.com/water-
feature-design/childrens-museum-sonoma-county

4-11

https://www.ryderarchitecture.com/project/royal-
stoke-university-hospital/

6-03

Screenshot from https://earth.google.com/web/

6-04

Screenshot from https://lbsyun.baidu.com/

6-05

Screenshot from https://www.qgis.org/en/site/

6-06

https://en.wikipedia.org/wiki/French_formal_garden

6-07

https://www.exeterinternational.com/blog/jewish-
berlin-6-essential-sights/

6-09

https://www.archdaily.cn/cn/967301/rang-gong-han-
huan-fa-xin-sheng-ling-ren-liang-tan-de-geng-xin-
xiang-mu-zheng-zai-zhong-gou-cheng-shi-sheng-huo

6-10

https://www.tclf.org/landscapes/noguchi-museum

6-11

https://japanobjects.com/features/garden-design

6-13

https://www.asla.org/2012awards/036.html

6-15

https://labgov.city/theurbanmedialab/will-urban-
farming-save-our-cities-perspectives-from-detroit-
and-brussels/

6-16

https://www.chicagotribune.com/opinion/editorials/
ct-edit-riverwalk-pollution-edit-0919-20150918-story.
html

6-21

https://www.archdaily.com/993614/abandoned-airport-near-athens-greece-set-to-be-transformed-into-europes-largest-coastal-park?ad_source=search&ad_medium=projects_tab&ad_source=search&ad_medium=search_result_all

7-01–7-05

https://www.gooood.cn/2021-asla-general-design-award-of-honor-the-leon-levy-native-plant-preserve-raymond-jungles-inc.htm

7-06–7-11

https://www.gooood.cn/2021-asla-research-award-of-honor-use-low-impact-development-facilities-to-build-an-ecological-sewage-treatment-system-for-remote-areas-gvl-design-group.htm

7-12, 7-13

http://www.landscape.cn/article/67147.html

7-16

http://www.landscape.cn/article/67147.html

7-17–7-20

http://www.landscape.cn/landscape/action/ShowInfo.php?classid=3&id=10029

7-21–7-25

https://www.gooood.cn/2019-asla-general-design-award-of-honor-hunters-point-south-waterfront-park-phase-ii-by-swabalsley-and-weissmanfredi-with-arup.htm

7-26, 7-27

https://www.gooood.cn/2021-asla-general-design-award-of-honor-duke-university-water-reclamation-pond-nelson-byrd-woltz-landscape-architects.htm

7-30

https://www.mecanoo.nl/Projects/project/233/Nieuw-Land-National-Park?d=0

7-31

https://mooool.com/en/forest-park-in-bad-lippspringe-by-sinai.html

7-32

https://www.sohu.com/a/533996938_121124642

7-33

https://www.archdaily.com/928594/why-does-venice-flood-and-what-is-being-done-about-it

7-34

https://www.sohu.com/a/533996938_121124642

7-35

https://julianamalta.com/domino-park-parque-novo-em-williamsburg-com-vista-para-manhattan/

7-37

https://www.sasaki.com/zh/projects/chicago-riverwalk/

7-38

https://en.wikipedia.org/wiki/Cheonggyecheon

7-39

https://www.metalocus.es/en/news/sea-a-public-space-aarhus-harbor-bath-big

7-40

https://www.ideabooom.com/7672

7-41

http://www.24901milesaround.com/blog/2016/5/12/day-9-foweyeden-project

7-42

https://playground-landscape.com/

7-43, 7-44

https://bbs.zhulong.com/101020_group_201864/detail32861754/

7-45

https://www.npr.org/2016/10/16/496865680/6-things-
you-need-to-know-about-cycling-on-the-sidewalk

7-46, 7-47

http://www.360doc.com/content/19/0218/02/
23036362_815764090.shtml

7-48

https://www.prnasia.com/story/204251-1.shtml

8-03

https://www.telegraph.co.uk/culture/
photography/11596966/Detroit-an-empty-city-from-
the-air.html

8-04

Provided by the Wu Penghan design team of East
China Normal University

8-06

https://ongreening.com/en/Resources/how-leed-
certification-work-1293

8-09

https://www.archdaily.com/885340/housing-
construction-in-argentina-uses-recycled-pet-bricks

8-11

https://www.archpaper.com/2020/03/google-arts-
culture-over-500-virtual-museums/

8-12

https://patkay.com/blogs/pk/teamlab-borderless-
tokyo

8-13

https://www.insider.com/video-virtual-wave-swirl-
huge-glass-box-screen-installation-seoul-2020-5

注释
Notes

[1]   Landscape gardening, the art of making gardens, parks, and around buildings look more natural and attractive. Cambridge dictionary, https://dictionary.cambridge.org/

[2]   Landscape architecture involves the planning, design, management, and nurturing of the built and natural environments. "What is landscape architecture？", American Society of Landscape Architects, https://www.asla.org/about

[3]   俞孔坚、李迪华：《〈景观设计：专业学科与教育〉导读》，《中国园林》2004年第5期。

[4]   张兵：《历史城镇整体保护中的"关联性"与"系统方法"——对"历史性城市景观"概念的观察和思考》，《城市规划》2014年第38期。

[5]   也称英式庭院，起源于18世纪的英国园艺造景，追求自然的景观美，由广大苑池构成自然风景式庭院。

[6]   弗雷德里克·奥姆斯特德，美国景观设计学奠基人，美国最重要的公园设计者。

[7]   卡尔弗·沃克斯，英裔美国建筑师，与奥姆斯特德一同参与设计了著名的美国纽约中央公园。

[8]   路德维希·密斯·凡德罗，德国建筑师，著名现代主义建筑大师。

[9]   克里斯托弗·唐纳德，出生于加拿大的著名景观设计师、花园设计师、城市规划师。

[10]   Tunnard, C. and Hunt, J. D. *Gardens in the Modern Landscape: A Facsimile of the Revised 1948 Edition*, University of Pennsylvania Press, 2014.

[11]   花园城市或称田园城市，是人类社区包围于田地或者花园的区域中，平衡住宅、工业和农业区域比例的一种城市规划。

[12]   王世福、易智康、张晓阳：《中国城市更新转型的反思与展望》，《城市规划学刊》2023年第1期。

[13]   [英]伊恩·汤普森：《景观设计学》，安聪译，译林出版社，2022年。

[14]   朱汉龙：《意大利后现代设计景观浅析》，《南京艺术学院学报（美术与设计）》2020年第3期。

[15]   安迪·高兹沃斯，英国雕塑家，摄影师和环保主义者，他创作了位于自然和城市环境特定地点的雕塑和大地艺术品。

[16]   林璎，美籍华裔建筑师，美国越战阵亡将士纪念碑、美国华人博物馆、公民权利纪念碑设计者。

[17]   安尼施·卡普尔，印度裔英国雕塑家，擅长装置艺术和观念艺术。著名雕塑作品《云门》《天空之镜》的创作者，作品被泰特美术馆和纽约现代艺术博物馆收藏。

[18]   Zev, N. and Lieberman, A. S. *Landscape Ecology: Theory and Application*. Springer-Verlag, 1984.

[19]   俞孔坚：《土地的设计:景观的科学与艺术》，《规划师》2004年第2期。

[20]   美国高线公园，位于美国纽约市曼哈顿废弃的纽约中央铁路西区线一个高架桥上的绿道和带状公园，长约2.33千米。

[21]   Baccarini C, Condon R, Eber H, et al.：《高线公园　美国纽约市》，《世界建筑导报》2021年第6期。

[22]   陈从周：《苏州园林》，同济大学出版社，2018年。

[23]   [美]莫森·莫斯塔法维、加雷斯·多尔蒂编著：《生态都市主义》，俞孔坚译，江苏科学技术出版社，2014年。

[24]   Yi-xi W. and Ying S. "The Role of the Landscape Architect in the 21st Century Fight against Climate Change." *International Journal of Liberal Arts and Social Science*, 2019, 7(10): 27-35.

[25]   王齐：《基于SITES体系的城市景观可持续设计研究》，天津科技大学2019年硕士学位

论文。

[26] Leadership in Energy and Environmental Design (LEED), 由非营利性的美国绿色建筑委员会开发，包括用于绿色建筑、房屋和社区的设计、建造、运营和维护的评级系统。

[27] Forest Stewardship Council（FSC），森林管理委员会，非营利性机构，由希望阻止森林遭到不断破坏的非政府机构、环保人士、木材贸易组织及具有社会责任感的消费者等构成。

[28] 张颢晖等：《移动滴灌系统土壤水分入渗试验与数值模拟》，《农业工程学报》2023年第6期。

[29] 吴隽宇、梁策：《风景园林视野下我国微气候研究概述与进展》，《南方建筑》2019年第6期。

[30] Suša, O. "Global dynamics of socio-environmental crisis: Dangers on the way to a sustainable future." *Civitas–Revista de Ciências Sociais*, 2019(19): 315–336.

[31] Seferlis, P, Varbanov P. S., Papadopoulos A. I., et al. "Sustainable design, integration, and operation for energy high-performance process systems." *Energy*, 2021(224): 120–158.

[32] 气候的特征，检索自科普中国。https://www.kepuchina.cn.

[33] 李丹宁、刘东云、王鑫：《缓解城市热岛效应的硬质景观设计方法研究综述》，《风景园林》2022年第8期。

[34] 翡翠项链，美国波士顿和布鲁克莱恩公园带的别称，占地约4.5平方千米。

[35] 李智兴、潘鑫晨、董庆鑫：《中低层住宅建筑形态气候适应性优化设计策略研究》，《工业建筑》2022年第7期。

[36] 近代骑楼是岭南传统民居与西方建筑艺术相结合演变而成的一种商住建筑形式，它是城市生态的一部分，记录着城市成长的过程，生动地折射出一个时代的人文风貌。叶曙明：《骑楼》，广东教育出版社，2010年。

[37] [美]约翰·O. 西蒙兹：《景观设计学》，俞孔坚等译，中国建筑工业出版社，2000年。

[38] [美]约翰·O. 西蒙兹：《景观设计学》，2000年。

[39] 狭管效应，也叫"峡谷效应"，指由于地形狭窄导致空气流速增大的现象。

[40] 冯娴慧：《城市的风环境效应与通风改善的规划途径分析》，《风景园林》2014年第5期。

[41] 雨水花园，用于汇聚和吸收屋顶或地面雨水的浅凹绿地。通过植物和沙土的综合净化作用，雨水可以起到涵养地下水或者补给城市用水的效果。

[42] 蔬菜园，这里特指欧洲中世纪的蔬菜园，以大型的草本花园为主，种植药草或香草。

[43] 枯山水，一般指由细沙碎石铺地，再加上一些叠放有致的石组所构成的缩微式园林景观，偶尔也包含苔藓、草坪或其他自然元素。

[44] 景天属植物，指一年生或多年生草本或亚灌木植物，有毛或无毛。

[45] 皮特·奥多夫，荷兰园艺设计师，"新多年生植物"运动的领军人物。他的设计大胆地使用多年生草本植物。

[46] [丹]斯坦·埃勒·拉斯穆森：《城镇与建筑》，韩煜译，天津大学出版社，2013年。

[47] 庞贝，古罗马城市，位于那不勒斯湾维苏威火山脚下，公元79年毁于维苏威火山大爆发。

[48] 梁陈：《隋唐长安都城区域山水人文空间格局营造研究》，陕西师范大学2022年博士学位论文。

[49] 街区是城市设计的重要元素。一个街区是四周由街道所围成的最小的区块，其内涵盖建地、建筑物等。

[50] 丝绸之路，一般指陆上丝绸之路，以古

代城市长安或洛阳为起点，经甘肃、新疆，到中亚、西亚，并连接地中海各国的陆上通道。

[51]　Gioielli, R. *Environmental Activism and the Urban Crisis: Baltimore, St. Louis, Chicago*. Temple University Press, 2014.

[52]　[美]查尔斯·瓦尔德海姆：《景观都市主义：从起源到演变》，陈崇贤、夏宇译，江苏凤凰科学技术出版社，2018年。

[53]　再生水，指废水或雨水经适当处理后，达到一定的水质指标，满足使用要求的水。

[54]　张翔、曾任之：《新加坡建筑立体绿化空间社会功能性研究》，《中外建筑》2022年第12期。

[55]　特雷维喷泉，俗称"许愿池"，是一座位于意大利罗马的喷泉，也是罗马最大的巴洛克风格喷泉。

[56]　郑曦：《城市蓝绿空间系统》，《风景园林》2022年第12期。

[57]　袁秋玲等：《食物—能源—水关联视角下蓝绿基础设施提升城市韧性的概念框架》，《城市发展研究》2022年第8期。

[58]　[美]尼古拉斯·T.丹尼斯、凯尔·D.布朗：《景观设计师便携手册》，刘玉杰、吉庆萍、俞孔坚译，中国建筑工业出版社，2002年。

[59]　张纯：《城市社区形态与再生》，东南大学出版社，2014年。

[60]　尼尔·G.柯克伍德、孙一鹤：《未来景观设计实践的维度》，《景观设计学》2014年第2期。

[61]　[美]约翰·O.西蒙兹：《景观设计学》，2000年。

[62]　王佳：《基于低影响开发的场地景观规划设计方法研究》，北京建筑大学2013年硕士学位论文。

[63]　赵婉彤等：《生活圈视角下的社区医疗设施点可达性研究——以长春市朝阳区为例》，《吉林建筑大学学报》2022年第5期。

[64]　华晨等：《社区商业设施空间步行可达性评价及布局优化——以绍兴市三区为例》，《浙江大学学报（工学版）》2022年第2期。

[65]　马赛公寓（法语Unité d'Habitation），意为"居住单元"或"居住统一体"，是由建筑师勒·柯布西耶设计的位于法国马赛的现代主义建筑。

[66]　树枝状线路，也称树枝状系统。该系统是希尔伯塞莫提出的，意在采用树枝状的道路系统将不同速度要求的车行交通及人行交通分开。

[67]　张刚：《城市公共设施的人性化设计思考》，《包装工程》2022年第20期。

[68]　谷康、黄丽江、杨艺红：《城市广场公共空间人性化设计研究》，东南大学出版社，2021年。

[69]　陈燕申、陈思凯：《丹麦哥本哈根市自行车发展战略探讨及启示》，《现代城市研究》2018年第2期。

[70]　王健：《大数据助力江苏智慧体育公园全民健身服务高质量发展研究》，南京体育学院2022年硕士学位论文。

[71]　陈锦赐：《以环境共生观营造共生城乡景观环境》，《城市发展研究》2004年第6期。

[72]　吴婉儿、黄春晓：《"自下而上"混合居住的老城社区社会空间特征研究——以苏州古城社区为例》，《建筑与文化》2020年第7期。

[73]　[美]克莱尔·库珀·马库斯、卡罗琳·弗朗西斯：《人性场所》，孙鹏译，中国建筑工业出版社，2001年。

[74]　超级线性公园，一个约804米长的城市空间，其所贯穿的区域是丹麦最具种族多样化、社会挑战性最强的街区之一。

[75]　Ellison, C. and Maynard, E. S. *Healing for the city: Counseling in the Urban Setting*. Wipf and Stock Publishers, 2002.

[76]　周详、常婧超：《城市治理与空间转型背景下上海遗产社区建设和公众参与机制研

究》，《现代城市研究》2023年第1期。

[77] 钟乐、杨锐：《国家公园定义比较研究》，《中华环境》2019年第8期。

[78] Oh, J. E. and Ma, H. "Enhancing visitor experience of theme park attractions: Focusing on animation and narrative." *Journal of Advanced Research in Dynamical and Control System*, 2018(4): 178–185.

[79] Afacan Y. "Impacts of urban living lab (ULL) on learning to design inclusive, sustainable, and climate-resilient urban environments." *Land Use Policy*, 2023(124): 106–443.

[80] [美]尼古拉斯·T. 丹尼斯、凯尔·D. 布朗：《景观设计师便携手册》，2002年。

[81] 朱建宁等：《西方园林史：19世纪之前》，中国林业出版社，2008年。

[82] 过渡性空间，又名"灰空间"，特指不割裂内外，又独立于内外的一个媒介结合区域。这个区域可提供室内与室外的中途点。

[83] 刘文忠、孙湘明：《障碍设计：实现人类平等与履行社会责任的设计智慧》，《艺术设计研究》2020年第4期。

[84] [美]梅格·卡尔金斯：《可持续景观设计：场地设计方法、策略与实践》，贾培义等译，中国建筑工业出版社，2016年。

[85] 刘悦来、谢宛芸：《共治的景观系列参与式设计营造工作坊——基于社区公共空间治理的景观教学模式融合探索》，《园林》2022年第12期。

[86] 乐颐生境花园将"生物环境"和"花园"结合在一起。该花园约732平方米，并成功入选"生态多样性100+全球典型案例"名单。

[87] Delbanco, A. ed. *Nature in Writing New England: An Anthology from the Puritans to the Present.* Harvard University Press, 2001.

[88] Birnbaum, Charles, ed. The Ford Foundation Project on The Cultural Landscape Foundation website, https://www.tclf.org/landscapes/ford-foundation-atrium.

[89] Remick, Rachel, ed. Ana Mendiata entry on the Museum of Modern Art website, https://womennart.com/2020/02/05/who-was-ana-mendieta/.

[90] Fernando-Galiano, L. ed., "Weekend House" in *A V Monograph 121 Sanaa: Sejima & Nishizawa*, 2006(8): 112–115.

[91] Lin, M. The Vietnam Veterans Memorial on website, https://www.mayalinstudio.com/memory-works/vietnam-veterans-memorial.

[92] Hine, H. Desert Song in ARTFORUM. https://www.artforum.com/features/desert-song-205228/.

[93] Anmahian, A. private conversation with author, May 2008.

[94] 1英尺约等于0.3米。

[95] 王志芳：《景观设计研究方法》，中国建筑工业出版社，2022年。

[96] [美]盖里·哈克、梁思思：《场地规划与设计（上）认知·方法》，梁思思译，中国建筑工业出版社，2022年。

[97] [美]约翰·O. 西蒙兹：《景观设计学》，2000年。

[98] [英]詹姆斯·布莱克：《景观与园林设计指南（原著第二版）》，张云路等译，中国建筑工业出版社，2020年。

[99] 赵春丽、杨滨章、刘岱宗：《PSPL调研法：城市公共空间和公共生活质量的评价方法——扬·盖尔城市公共空间设计理论与方法探析（3）》，《中国园林》2012年第9期。

[100] 王扬：《韧性理念下的"建筑-场地"设计思考》，《建筑与文化》2023年第4期。

[101] Ouyang, P. and Wu, X. "Analysis and Evaluation of the Service Capacity of a Waterfront Public Space Using Point-of-Interest Data Combined with Questionnaire Surveys ." *Land*, 2013 (7).

[102] 张琴、孙晓珂：《"景观设计原理"课程中的场地数据调查与分析教学研究》，《创意与设计》2023年第2期。

[103] 孟彤：《景观元素设计理论与方法》，中国建筑工业出版社，2012年。

[104] 张德顺、孙力、Marie Simon：《法国园林发展的三个时代》，《上海交通大学学报（农业科学版）》2019年第4期。

[105] 法国圣殿广场原是19世纪末的巨大钢铁结构传统建筑，于2007年由法国Milou建筑工作室完成更新改造。在2014年重新向公众开放。

[106] 邹涵、刘书君：《基于文化生态学的城市历史环境色彩量化研究——以武汉市汉正街为例》，《城市建筑》2022年第19期。

[107] 郭敏：《江南园林声景主观评价及设计策略》，浙江大学2014年博士学位论文。

[108] 加拿大糖果海滩，由加拿大设计事务所Claude Cormier + Associés设计。该项目与2010年建成，约8500平方米。

[109] 卢韵琴等：《〈日本香景100选〉的解读与拓展》，《南方建筑》2022年第10期。

[110] 张亚丽等：《案例分析法在〈城市规划原理〉教学中的应用——以非城市规划专业为例》，《安徽农学通报（上半月刊）》，2009年第1期。

[111] 庄启璇：《基于GIS分析的沿海社区弹性景观设计研究——以纳格斯海德（Nags Head）为例》，《建筑与文化》2022年第12期。

[112] 李慧希：《基于地图术（Mapping）的景观建筑学理论研究》，东南大学2016年博士学位论文。

[113] 韩林桅、张淼、石龙宇：《生态基础设施的定义、内涵及其服务能力研究进展》，《生态学报》2019年第19期。

[114] 沈阳应用生态研究所：《沈阳生态所等在城市景观调节生态气候效应方面获进展》，《高科技与产业化》2022年第5期。

[115] 马辉、刘文欣、潘宥承：《东北少数民族景观建筑低成本创作实践》，《艺术工作》2022年第5期。

[116] BSLA Senior Studio. "The Emerald Network: Connecting & Extending Boston's Greenways." *Adapting to Expanding and Contracting Cities*, 2019: 174.

[117] 章明、张洁、秦曙：《风景的媒介——杨浦滨江雨水花园的四重叙事》，《中国园林》2021年第7期。

[118] 孙群郎、夏英华：《美国巴尔的摩的棕地环境治理与再开发》，《历史教学问题》2020年第6期。

[119] 邹锦：《城市滨水空间的韧性机理及其设计响应》，《上海城市规划》2023年第1期。

[120] 闫婧等：《不同气候区绿色屋顶蒸散发模拟研究》，《生态学报》2023年第43期。

[121] Dalley, S. *The Mystery of the Hanging Gardens of Babylon. Elusive World Wonder Traced*. OUP Oxford, 2013; Finkel, I. L. and Seymour, M. J. ed. *Babylon: Myth and Reality*. British Museum Press, 2008; Cotterell, A. *The First Great Powers: Babylon and Assyria*. Hurst, 2019.

[122] 海绵城市，新一代城市雨洪管理概念，指城市能够像海绵一样，在适应环境变化和应对雨水带来的自然灾害等方面具有良好的弹性，也称为"水弹性城市"。

[123] 缘起美国纽约2012年飓风桑迪带来的破坏而由政府举办的海岸保护计划。

[124] 章晶晶等：《基于关联规则的儿童户外活动空间偏好研究——以杭州三个社区公园为例》，《中国园林》2023年第5期。

[125] 翁锡全等：《城市建筑环境对居民身体活动和健康的影响——运动与健康促进研究新领域》，《体育科学》2010年第9期。

[126] [美]约翰·L.摩特洛克：《景观设计导论（第二版）》，于矛译，天津大学出版社，

2016年。

[127] 俞孔坚：《大历史视野中的人类景观》，《景观设计学（中英文）》2021年第2期。

[128] [美]查尔斯·瓦尔德海姆：《景观都市主义：从起源到演变》，2018年。

[129] Corner, J. and Hirsch, A. B. ed. *The landscape imageination: Collected Essays of James Corner 1990–2010.* 2014.

[130] "都市针灸"概念最初源自西班牙建筑师和城市学家马拉勒斯，意指用类似"中医针灸"的方法，对城市的"疾病"进行治愈。以小空间、巧方法、低投入代替大拆大改的改造方式，提升城市的品质。

[131] 俞孔坚、李迪华：《可持续景观》，《城市环境设计》2007年第1期。

[132] Abbas, K., and Sohair Kadhem Abd. "Study of using of recycled brick waste (RBW) to produce environmental friendly concrete: A review." *Journal of Engineering*, 2021(27): 1–14.

[133] 高娣、王龙意：《数字技术在园林景观设计中的表达与应用》，《北京规划建设》2022年第4期。

[134] 石义金等：《公众认知视角下的"印象·城市"智慧服务平台构建研究》，《现代情报》2021年第3期。

[135] 武静等：《基于生态智慧的武汉凤凰湖公园规划研究》，《西北师范大学学报（自然科学版）》2022年第2期。

[136] 饶永、李新宇：《数字技术时代体验性博物馆展示设计研究》，《家具与室内装饰》2021年第12期。

YOLANDA MARKET
优康·和集

232

OLD FARMERS MARKET
YOLANDA MARKET
优康·和集

KUNEKUNE

参考资料
References

著作

1. [丹]斯坦·埃勒·拉斯穆森：《城镇与建筑》，韩煜译，天津大学出版社，2013年。

2. [荷]雷姆·库哈斯：《癫狂的纽约》，唐克扬译，三联书店，2015年。

3. [美]查尔斯·瓦尔德海姆：《景观都市主义：从起源到演变》，陈崇贤、夏宇译，江苏凤凰科学技术出版社，2018年。

4. [美]盖里·哈克、梁思思：《场地规划与设计（上）认知·方法》，梁思思译，中国建筑工业出版社，2022年。

5. [美]克莱尔·库珀·马库斯、卡罗琳·弗朗西斯：《人性场所》，孙鹏译，中国建筑工业出版社，2001年。

6. [美]梅格·卡尔金斯：《可持续景观设计：场地设计方法、策略与实践》，贾培义等译，中国建筑工业出版社，2016年。

7. [美]莫森·莫斯塔法维、加雷斯·多尔蒂编著：《生态都市主义》，俞孔坚译，江苏科学技术出版社，2014年。

8. [美]尼古拉斯·T.丹尼斯、凯尔·D.布朗：《景观设计师便携手册》，刘玉杰、吉庆萍、俞孔坚译，中国建筑工业出版社，2002年。

9. [美]约翰·L.摩特洛克：《景观设计导论（第二版）》，于矛译，天津大学出版社，2016年。

10. [美]约翰·O.西蒙兹：《景观设计学》，俞孔坚等译，中国建筑工业出版社，2000年。

11. [英]伊恩·汤普森：《景观设计学》，安聪译，译林出版社，2022年。

12. [英]詹姆斯·布莱克：《景观与园林设计指南（原著第二版）》，张云路等译，中国建筑工业出版社，2020年。

13. SWA基础设施研究提案编：《景观基础设施：SWA事务所作品分析（原著第二版）》，曾颖译，中国建筑工业出版社，2014年。

14. 陈从周：《苏州园林》，同济大学出版社，2018年。

15. 成玉宁：《数字景观》，东南大学出版社，2019年。

16. 冯璐：《弹性景观：风暴潮适应性景观基础设施》，东南大学出版社，2020年。

17. 谷康、黄丽江、杨艺红：《城市广场公共空间人性化设计研究》，东南大学出版社，2021年。

18. 孟彤：《景观元素设计理论与方法》，中国建筑工业出版社，2012年。

19. 王志芳：《景观设计研究方法》，中国建筑工业出版社，2022年。

20. 叶曙明：《骑楼》，广东教育出版社，2010年。

21. 张纯：《城市社区形态与再生》，东南大学出版社，2014年。

22. 朱建宁等：《西方园林史：19世纪之前》，中国林业出版社，2008年。

期刊论文

1. Baccarini C, Condon R, Eber H, et al.：《高线公园 美国纽约市》，《世界建筑导报》2021年第6期。

2. 陈锦赐：《以环境共生观营造共生城乡景观环境》，《城市发展研究》2004年第6期。

3. 陈燕申、陈思凯：《丹麦哥本哈根市自行车发展战略探讨及启示》，《现代城市研究》2018年第2期。

4. 冯娴慧：《城市的风环境效应与通风改善的规划途径分析》，《风景园林》2014年第5期。

5. 高娣、王龙意：《数字技术在园林景观设计中的表达与应用》，《北京规划建设》2022年第4期。

6. 韩林桅、张淼、石龙宇：《生态基础设施的定义、内涵及其服务能力研究进展》，《生态学报》2019年第19期。

7. 何志森：《Mapping工作坊：重新解读城市更新与日常生活的关系》，《景观设计学》2017年第5期。

8. 何志森、杨薇芬：《大地之上：基于人的尺度的图绘》，《国际城市规划》2019年第6期。

9. 华晨等：《社区商业设施空间步行可达性评价及布局优化——以绍兴市三区为例》，《浙江大学学报（工学版）》2022年第2期。

10. 李丹宁、刘东云、王鑫：《缓解城市热岛效应的硬质景观设计方法研究综述》，《风景园林》2022年第8期。

11. 李智兴、潘鑫晨、董庆鑫：《中低层住宅建筑形态气候适应性优化设计策略研究》，《工业建筑》2022年第7期。

12. 刘文忠、孙湘明：《障碍设计：实现人类平等与履行社会责任的设计智慧》，《艺术设计研究》2020年第4期。

13. 刘悦来、谢宛芸：《共治的景观系列参与式设计营造工作坊——基于社区公共空间治理的景观教学模式融合探索》，《园林》2022年第12期。

14. 卢韵琴等：《〈日本香景100选〉的解读与拓展》，《南方建筑》2022年第10期。

15. 马辉、刘文欣、潘宥承：《东北少数民族景观建筑低成本创作实践》，《艺术工作》2022年第5期。

16. 尼尔·G. 柯克伍德、孙一鹤：《未来景观设计实践的维度》，《景观设计学》2014年第2期。

17. 饶永、李新宇：《数字技术时代体验性博物馆展示设计研究》，《家具与室内装饰》2021年第12期。

18. 沈阳应用生态研究所：《沈阳生态所等在城市景观调节生态气候效应方面获进展》，《高科技与产业化》2022年第5期。

19. 石义金等：《公众认知视角下的"印象·城市"智慧服务平台构建研究》，《现代情报》2021年第3期。

20. 孙群郎、夏英华：《美国巴尔的摩的棕地环境治理与再开发》，《历史教学问题》2020年第6期。

21. 王世福、易智康、张晓阳：《中国城市更新转型的反思与展望》，《城市规划学刊》2023年第1期。

22. 王扬：《韧性理念下的"建筑-场地"设计思考》，《建筑与文化》2023年第4期。

23. 翁锡全等：《城市建筑环境对居民身体活动和健康的影响——运动与健康促进研究新领域》，《体育科学》2010年第9期。

24. 吴隽宇、梁策：《风景园林视野下我国微气候研究概述与进展》，《南方建筑》2019年第6期。

25. 吴婉儿、黄春晓：《"自下而上"混合居住的老城社区社会空间特征研究——以苏州古城社区为例》，《建筑与文化》2020年第7期。

26. 武静等：《基于生态智慧的武汉凤凰湖公园规划研究》，《西北师范大学学报（自然科学版）》2022年第2期。

27.闫婧等：《不同气候区绿色屋顶蒸散发模拟研究》，《生态学报》2023年第43期。

28.俞孔坚、李迪华：《〈景观设计：专业学科与教育〉导读》，《中国园林》2004年第5期。

29.俞孔坚、李迪华：《可持续景观》，《城市环境设计》2007年第1期。

30.俞孔坚：《大历史视野中的人类景观》，《景观设计学（中英文）》2021年第2期。

31.俞孔坚：《土地的设计:景观的科学与艺术》，《规划师》2004年第2期。

32.袁秋玲等：《食物—能源—水关联视角下蓝绿基础设施提升城市韧性的概念框架》，《城市发展研究》2022年第8期。

33.张兵：《历史城镇整体保护中的"关联性"与"系统方法"——对"历史性城市景观"概念的观察和思考》，《城市规划》2014年第38期。

34.张德顺、孙力、Marie Simon：《法国园林发展的三个时代》，《上海交通大学学报（农业科学版）》2019年第4期。

35.张刚：《城市公共设施的人性化设计思考》，《包装工程》2022年第20期。

36.张颢晖等：《移动滴灌系统土壤水分入渗试验与数值模拟》，《农业工程学报》2023年第6期。

37.张琴、孙晓珂：《"景观设计原理"课程中的场地数据调查与分析教学研究》，《创意与设计》2023年第2期。

38.张翔、曾任之：《新加坡建筑立体绿化空间社会功能性研究》，《中外建筑》2022年第12期。

39.张亚丽等：《案例分析法在〈城市规划原理〉教学中的应用——以非城市规划专业为例》，《安徽农学通报（上半月刊）》，2009年第1期。

40.章晶晶等：《基于关联规则的儿童户外活动空间偏好研究——以杭州三个社区公园为例》，《中国园林》2023年第5期。

41.章明、张洁、秦曙：《风景的媒介——杨浦滨江雨水花园的四重叙事》，《中国园林》2021年第7期。

42.赵春丽、杨滨章、刘岱宗：《PSPL调研法：城市公共空间和公共生活质量的评价方法——扬·盖尔城市公共空间设计理论与方法探析（3）》，《中国园林》2012年第9期。

43.赵婉彤等：《生活圈视角下的社区医疗设施点可达性研究——以长春市朝阳区为例》，《吉林建筑大学学报》2022年第5期。

44.郑曦：《城市蓝绿空间系统》，《风景园林》2022年第12期。

45.钟乐、杨锐：《国家公园定义比较研究》，《中华环境》2019年第8期。

46.周详、常婧超：《城市治理与空间转型背景下上海遗产社区建设和公众参与机制研究》，《现代城市研究》2023年第1期。

47.朱汉龙：《意大利后现代设计景观浅析》，《南京艺术学院学报（美术与设计）》2020年第3期。

48.庄启璇：《基于GIS分析的沿海社区弹性景观设计研究——以纳格斯海德（Nags Head）为例》，《建筑与文化》2022年第12期。

49.邹涵、刘书君：《基于文化生态学的城市历史环境色彩量化研究——以武汉市汉正街为例》，《城市建筑》2022年第19期。

50.邹锦：《城市滨水空间的韧性机理及其设计响应》，《上海城市规划》2023年第1期。

学位论文

1. 郭敏：《江南园林声景主观评价及设计策略》，浙江大学2014年博士学位论文。
2. 李慧希：《基于地图术（Mapping）的景观建筑学理论研究》，东南大学2016年博士学位论文。
3. 梁陈：《隋唐长安都城区域山水人文空间格局营造研究》，陕西师范大学2022年博士学位论文。
4. 王佳：《基于低影响开发的场地景观规划设计方法研究》，北京建筑大学2013年硕士学位论文。
5. 王健：《大数据助力江苏智慧体育公园全民健身服务高质量发展研究》，南京体育学院2022年硕士学位论文。
6. 王齐：《基于SITES体系的城市景观可持续设计研究》，天津科技大学2019年硕士学位论文。

英文文献

1. Abbas, K., and Sohair Kadhem Abd. "Study of using of recycled brick waste (RBW) to produce environmental friendly concrete: A review." *Journal of Engineering*, 2021(27): 1–14.
2. Afacan Y. "Impacts of urban living lab (ULL) on learning to design inclusive, sustainable, and climate-resilient urban environments." *Land Use Policy*, 2023(124): 106–443.
3. Barber, D. A. *Modern Architecture and Climate: Design before Air Conditioning*. Princeton University Press, 2023.
4. Bridge, G. "The Hole World: Scales and Spaces of Extraction." *Scenario 05: Extraction*, Fall 2015, viewed online.
5. BSLA Senior Studio. "The Emerald Network: Connecting & Extending Boston's Greenways." *Adapting to Expanding and Contracting Cities*, 2019: 174.
6. Cheshmehzangi, A. *Green Infrastructure in Chinese Cities*. Springer, 2022.
7. Cook, D. I., U.S. *Forest Service, and Van Haverbeke D. F. Trees and Shrubs for Noise Abatement*. University Press of the Pacific, 2004.
8. Corner, J. and Hirsch, A. B. ed. *The landscape imageination: Collected Essays of James Corner 1990–2010*. 2014.
9. Cotterell, A. *The First Great Powers: Babylon and Assyria*. Hurst, 2019.
10. Cuff, D., et al. *Urban humanities: New practices for reimagining the city*. MIT Press, 2020.
11. Dalley, S. *The Mystery of the Hanging Gardens of Babylon. Elusive World Wonder Traced*. OUP Oxford, 2013.
12. Dee, C. *Form and Fabric in Landscape Architecture: A Visual Introduction*. Taylor & Francis, 2004.

13. Dekay, M. and Brown, G. Z. Sun, *Wind, and Light: Architectural Design Strategies*. John Wiley & Sons, 2013.

14. Delbanco, A. ed. *Nature in Writing New England: An Anthology from the Puritans to the Present*. Harvard University Press, 2001.

15. Di Mari, A. and Yoo, N. "Operative design." *Amsterdam: BIS*, 2013.

16. Eckbo, G. *Landscapes for Living*. Duell Sloan and Pearce, 1950, especially pp. 61–74.

17. Ellison, C. and Maynard, E. S. *Healing for the City: Counseling in the Urban Setting*. Wipf and Stock Publishers, 2002.

18. Farr, D. *Sustainable urbanism: Urban design with nature*. John Wiley & Sons, 2011.

19. Fernando–Galiano, L. ed., "Weekend House." *AV Monograph 121 Sanaa: Sejima & Nishizawa*, 2006(8): 112–115.

20. Finkel, I. L. and Seymour, M. J. ed. *Babylon: Myth and Reality*. British Museum Press, 2008.

21. Firehock, K. E. and Walker, R. A. *Green Infrastructure: Map and Plan the Natural World with GIS*. Esri Press, 2019.

22. Foster, S. R. and Iaione, C. *Co–Cities: Innovative Transitions toward Just and Self–Sustaining Communities*. The MIT Press, 2022.

23. Gehl, J. *Life Between Buildings: Using Public Space*. Island Press, 2011.

24. Giedion, S. *Space, time and architecture: the growth of a new tradition*. Harvard University Press, 2009.

25. Gioielli, R. *Environmental Activism and the Urban Crisis: Baltimore, St. Louis, Chicago*. Temple University Press, 2014.

26. Givoni, B. Man, *Climate and Architecture*. Van Nostrand Reinhold, 1981.

27. Goh, K., et al. *Just Urban Design: The Struggle for a Public City*. The MIT Press, 2022.

28. Hunt, J. D. *Greater perfections: the practice of garden theory*. University of Pennsylvania Press, 2000.

29. Kimmelman, M. *The Intimate City: Walking New York*. Penguin Press, 2022.

30. Kopec, D. *Environmental Psychology for Design*. Fairchild Books, 2018.

31. Lewis, P., Tsutumaki, M. and Lewis, D. J. *Manual of section*. Chronicle Books, 2016.

32. Litscher, M. "Jane Jacobs: the death and life of great American cities." *Schlüsselwerke der Stadtforschung*, 2017: 367–394.

33. LLynch, K. *The Image of the City*. MIT Press, 1964.

34. Manshel, A. M. *Learning from Bryant Park: Revitalizing Cities, Towns, and Public Spaces*. Rutgers University Press, 2020.

35. McHarg, I. L. and American Museum of Natural History. *Design with nature*. American Museum of Natural History, 1969.

36. Moscow, K. and Linn, R. S. *Small Scale: Creative Solutions for Better City Living*. Princeton Architectural Press, 2010.

37. Nakamura, F. *Green Infrastructure and Climate Change Adaptation: Function, Implementation and Governance*. Springer Nature, 2022.

38. Oh, J. E. and Ma, H. "Enhancing visitor experience of theme park attractions: Focusing on animation and narrative." *Journal of Advanced Research in Dynamical and Control System*, 2018(4): 178–185.

39. Olgyay, V. *Design with Climate: Bioclimatic Approach to Architectural Regionalism*. Princeton University Press, 2015.

40. Ouyang, P. and Wu, X. "Analysis and Evaluation of the Service Capacity of a Waterfront Public Space Using Point–of–Interest Data Combined with Questionnaire Surveys." *Land*, 2013(7).

41. Paez, R. *Operative Mapping: The Use of Maps as a Design Tool*. Actar, 2019.

42. Rasmussen, S E. *Experiencing Architecture* (Vol.2). MIT Press, 1964.

43. Reed, C. *Retooling Metropolis: Working Landscapes, Emergent Urbanism*. Harvard University Graduate School of Design, 2017.

44. Roesler, S. *City, Climate, and Architecture: A Theory of Collective Practice*. DeGruyter, 2022.

45. Seferlis, P, Varbanov P. S., Papadopoulos A. I., et al. "Sustainable design, integration, and operation for energy high−performance process systems." *Energy*, 2021(224): 120−158.

46. Sim, D. *Soft City: Building Density for Everyday Life*. Island Press, 2019.

47. Simitch, A. and Warke, V. *The Language of Architecture: 26 Principles Every Architect Should Know*. Rockport Publishers, 2014, especially pp. 109−116.

48. Speck, J. *Walkable City: How Downtown Can Save America, One Step at a Time*. Macmilan, 2013.

49. Stahlschmidt, P., Swaffield, S. and et al. *Landscape Analysis: Investigating the potentials of space and place*. Routledge, 2017.

50. Starke, B.W. and Simonds, J.O. *Landscape Architecture: A Manual of Environmental Planning and Design*. McGraw−Hill, 2013.

51. Steg, L. and De Groot, I. M. *Environmental Psychology: An Introduction*. 2018.

52. Suša, O. "Global dynamics of socio−environmental crisis: Dangers on the way to a sustainable future." *Civitas−Revista de Ciências Sociais*, 2019(19): 315−336.

53. Thompson, I. *Landscape Architecture: A Very Short Introduction*. Oxford University Press, 2014.

54. Tunnard, C. and Hunt, J. D. *Gardens in the Modern Landscape: A Facsimile of the Revised 1948 Edition*, University of Pennsylvania Press, 2014.

55. Vidal, J. M. ed. "Pequeña escala = small scale." Special Issue, *Paisea*, 2014(28).

56. Waldheim, C. *Landscape as Urbanism: A General Theory*. Princeton University Press, 2016.

57. Waldheim, C. *The Landscape Urbanism Reader*. Princeton Architectural Press, 2006.

58. Weller, R., Drozdz, Z. and Kjaersgaard, S. P. "Hotspot cities: Identifying peri−urban conflict zones" *Journal of Landscape Architecture*, 2019(14): 8−19.

59. Whyte, W. H. *The Social Life of Small Urban Spaces*. Project for Public Spaces, 2001.

60. Yi−xi W. and Ying S. "The Role of the Landscape Architect in the 21st Century Fight against Climate Change." *International Journal of Liberal Arts and Social Science*, 2019, 7(10): 27−35.

61. Zev, N. and Lieberman, A. S. *Landscape Ecology: Theory and Application*. Springer−Verlag, 1984.

网络资料

1. Birnbaum, Charles, ed. The Ford Foundation Project on The Cultural Landscape Foundation website, https://www.tclf.org/landscapes/ford-foundation-atrium.
2. Cambridge dictionary, https://dictionary.cambridge.org/.
3. Hine, H. Desert Song in ARTFORUM. https://www.artforum.com/features/desert-song-205228/.
4. Landscape architecture involves the planning, design, management, and nurturing of the built and natural environments. "What is landscape architecture? ", American Society of Landscape Architects, https://www.asla.org/about.
5. Lin, M. The Vietnam Veterans Memorial on website, https://www.mayalinstudio.com/memory-works/vietnam-veterans-memorial.
6. Remick, Rachel, ed. Ana Mendiata entry on the Museum of Modern Art website, https://www.moma.org/artists/3924.
7. 科普中国，https://www.kepuchina.cn.
8. 联合国可持续发展目标章程，https://www.un.org/sustainabledevelopment/.
9. 苏州园林绿化管理局，http://ylj.suzhou.gov.cn.

后　记

　　本书的写作和出版得到了海内外老师及学生的大力支持，在此表示无限感激。首先，我很感激凯伦·尼尔森教授的无私帮助。我作为学生在美国留学期间，从尼尔森教授的景观设计理论课程中获得了很大的启发。其中重要的原因是，她不仅教授理论，而且开拓了我对景观设计的不同角度的思考，让我觉得"一切可为设计"。这也是本书章节的编写理念，并不是导向一个统一的标准解释，而是结合主题讨论拓展学生对于景观设计的认知。同时，这种小小的热情在我心里播下了种子，希望有一天我也可以把这种理念展示到国内的课堂上。作为教育从业者，尼尔森教授不仅帮助我修订了本书的英语章节，还为我引荐了美国波士顿建筑学院的学者，包括玛利亚·贝拉尔塔教授、马修教授等。其中，玛利亚·贝拉尔塔教授不仅来访上海参加两校专业合作的会议，还积极参与了景观设计课程的线上线下评图联动。他们对课程和教育交流的支持，为华东师范大学设计学院的学生提供了景观设计领域海外交流的机会，为两校长期的教育研究合作提供了稳定的基础。

　　另外，我还要感谢华东师范大学设计学院的魏劭农院长、沈榆老师、王锋老师等在教材建设过程中给予的支持，以及广西师范大学出版社对出品控制的严谨态度。正是因为设计学院与出版社共同构建的这一开明而富有创新精神的支持平台，才使得教材建设得以在自由的学术空间中实现创新。最后，我要感谢课程所有的历届学生，课堂上讨论的内容很多成为本书的素材以及主题讨论的灵感来源，也为我的课程提供了许多宝贵的反馈。特别致谢顾昕晔、欧阳品月、张雨珊，她们利用课余和寒假休息的时间积极参与了教材的编校工作。其中，顾昕晔负责了教材的整体版面设计和章节校对工作；欧阳品月负责了教材的版面设计和统筹工作；张雨珊不仅负责文字校对，还兼顾了章节的图示校对、参考文献检阅等烦琐的工作。再次感谢你们的加入，让这本书的呈现更为完美。

"景观设计导论"是一门需要不断考究经典，又需要不断更新探索的课程。本书内容多为笔者在海外学习、工作，以及近几年在教学和研究实践中形成的思路与理念。本书的图文内容涉及面较广，并且中英双语的呈现为本书的编写提供了较大的难度。书中难免出现研究的不足之处，恳请读者朋友批评指正，笔者后续会不断改进更新。

伍晓雯

2024 年 12 月于上海

Postscript

I want to convey my sincere gratitude to the instructors and students at home and abroad who have generously supported the writing and publication of this book. Before anything else, I want to thank Professor Karen Nelson for her constant support and generosity. While studying abroad in the United States, Professor Nelson's landscape design theory course greatly influenced me as a student. One of the reasons is that she not only taught landscape design theory but also exposed me to various landscape design perspectives and made me believe that "everything can be designed." It guides me to organize the book's chapters through topical discussion in order to expand the students' understanding of landscape architecture.

Likewise, this small passion has planted the seed in my mind that I aspire to present this concept in my Chinese classroom one day. As an educator, Professor Nelson contributed a chapter to the book and assisted me in editing the English manuscript. She also introduced me to faculty members at the Boston Architectural College, such as President Dr. Mahesh Daas, Professor Maria Bellata, and others. In addition to traveling to Shanghai to attend the professional collaboration meeting between the two institutions, Professor Maria Bellata actively participated in the online and offline evaluation of the landscape architecture studios. East China Normal University's School of Design students could join landscape architecture exchange programs abroad by supporting the course and educational exchanges. It has created a solid basis for long-term collaboration between the two universities in curriculum design and project research.

Furthermore, I want to thank Dean Shaonong Wei, Professor Yu Shen, and Associate Professor Feng Wang of the School of Design at East China Normal University for their help with the book-building procedure. With the support of the School of Design, an innovative and enlightened platform, the writing and creation of the book would have had academic autonomy and the school's encouragement for originality.

Lastly, I would like to thank all the students in the course, as many classroom discussions became material for this book and inspiration for the thematic discussions, in addition to providing me with a great deal of helpful feedback for the course. Special thanks to students Xinye Gu, Pinyue Ouyang, and Yushan Zhang. Thank you for joining us and contributing to the perfection of this book.

Introduction to Landscape Architecture is a course that demands a constant examination of the classics and continuous exploration. Many of the ideas and concepts expressed in this book were shaped by my recent study and employment abroad, as well as my teaching and research. The vast amount of graphic content in this book and the bilingual presentation of the Chinese and English versions made it more challenging to prepare.The findings of this study reveal certain limitations, leaving room for further exploration and refinement in our future endeavors within this field.

Xiaowen Wu

图书在版编目（CIP）数据

景观设计导论／伍晓雯，（美）凯伦·尼尔森著.
桂林：广西师范大学出版社，2025. 6. -- ISBN 978-
7-5598-7845-8

Ⅰ. TU983

中国国家版本馆 CIP 数据核字第 20253DG503 号

景观设计导论
JINGGUAN SHEJI DAOLUN

出 品 人：刘广汉
责任编辑：肖　莉
助理编辑：茹婧羽
装帧设计：马韵蕾　顾昕晔

广西师范大学出版社出版发行

（ 广西桂林市五里店路 9 号　　　　邮政编码：541004 ）
（ 网址：http://www.bbtpress.com ）

出版人：黄轩庄
全国新华书店经销
销售热线：021 - 65200318　021 - 31260822 - 898
山东临沂新华印刷物流集团有限责任公司印刷
（临沂高新技术产业开发区新华路 1 号 邮政编码：276017）
开本：787 mm × 1 092 mm　　1/16
印张：16.5　　　　　　　　字数：298 千
2025 年 6 月第 1 版　　　2025 年 6 月第 1 次印刷
定价：98.00 元

如发现印装质量问题，影响阅读，请与出版社发行部门联系调换。